Proofs Without Words III

Further Exercises in Visual Thinking

© 2015 by The Mathematical Association of America, Inc.

Library of Congress Catalog Card Number 2015955515

Print edition ISBN 978-0-88385-790-8
Electronic edition ISBN 978-1-61444-121-2

Printed in the United States of America

Current Printing (last digit):
10 9 8 7 6 5 4 3

Proofs Without Words III

Further Exercises in Visual Thinking

Roger B. Nelsen
Lewis & Clark College

Published and Distributed by

The Mathematical Association of America

CLASSROOM RESOURCE MATERIALS

Classroom Resource Materials is intended to provide supplementary classroom material for students—laboratory exercises, projects, historical information, textbooks with unusual approaches for presenting mathematical ideas, career information, etc.

101 Careers in Mathematics, 3rd edition edited by Andrew Sterrett

Archimedes: What Did He Do Besides Cry Eureka?, Sherman Stein

Arithmetical Wonderland, Andrew C. F. Liu

Calculus: An Active Approach with Projects, Stephen Hilbert, Diane Driscoll Schwartz, Stan Seltzer, John Maceli, and Eric Robinson

Calculus Mysteries and Thrillers, R. Grant Woods

Cameos for Calculus: Visualization in the First-Year Course, Roger B. Nelsen

Conjecture and Proof, Miklós Laczkovich

Counterexamples in Calculus, Sergiy Klymchuk

Creative Mathematics, H. S. Wall

Environmental Mathematics in the Classroom, edited by B. A. Fusaro and P. C. Kenschaft

Excursions in Classical Analysis: Pathways to Advanced Problem Solving and Undergraduate Research, by Hongwei Chen

Explorations in Complex Analysis, Michael A. Brilleslyper, Michael J. Dorff, Jane M. McDougall, James S. Rolf, Lisbeth E. Schaubroeck, Richard L. Stankewitz, and Kenneth Stephenson

Exploratory Examples for Real Analysis, Joanne E. Snow and Kirk E. Weller

Exploring Advanced Euclidean Geometry with GeoGebra, Gerard A. Venema

Game Theory Through Examples, Erich Prisner

Geometry From Africa: Mathematical and Educational Explorations, Paulus Gerdes

The Heart of Calculus: Explorations and Applications, Philip Anselone and John Lee

Historical Modules for the Teaching and Learning of Mathematics (CD), edited by Victor Katz and Karen Dee Michalowicz

Identification Numbers and Check Digit Schemes, Joseph Kirtland

Interdisciplinary Lively Application Projects, edited by Chris Arney

Inverse Problems: Activities for Undergraduates, Charles W. Groetsch

Keeping it R.E.A.L.: Research Experiences for All Learners, Carla D. Martin and Anthony Tongen

Laboratory Experiences in Group Theory, Ellen Maycock Parker

Learn from the Masters, Frank Swetz, John Fauvel, Otto Bekken, Bengt Johansson, and Victor Katz

Math Made Visual: Creating Images for Understanding Mathematics, Claudi Alsina and Roger B. Nelsen

Mathematics Galore!: The First Five Years of the St. Marks Institute of Mathematics, James Tanton

Methods for Euclidean Geometry, Owen Byer, Felix Lazebnik, and Deirdre L. Smeltzer

Ordinary Differential Equations: A Brief Eclectic Tour, David A. Sánchez

MAA Service Center
P.O. Box 91112
Washington, DC 20090-1112
1-800-331-1MAA   FAX: 1-301-206-9789

Introduction

A dull proof can be supplemented by a geometric analogue so simple and beautiful that the truth of a theorem is almost seen at a glance.

—Martin Gardner

About a year after the publication of *Proofs Without Words: Exercises in Visual Thinking* by the Mathematical Association of America in 1993, William Dunham, in his delightful book *The Mathematical Universe, An Alphabetical Journey through the Great Proofs, Problems, and Personalities* (John Wiley & Sons, New York, 1994), wrote

> Mathematicians admire proofs that are ingenious. But mathematicians especially admire proofs that are ingenious and economical—lean, spare arguments that cut directly to the heart of the matter and achieve their objectives with a striking immediacy. Such proofs are said to be elegant.
>
> Mathematical elegance is not unlike that of other creative enterprises. It has much in common with the artistic elegance of a Monet canvas that depicts a French landscape with a few deft brushstrokes or a haiku poem that says more than its words. Elegance is ultimately an aesthetic, not a mathematical property.
>
> … an ultimate elegance is achieved by what mathematicians call a "proof without words," in which a brilliantly conceived diagram conveys a proof instantly, without need even for explanation. It is hard to get more elegant than that.

Since the books mentioned above were published, a second collection *Proofs Without Words II: More Exercises in Visual Thinking* was published by the MAA in 2000, and this book constitutes the third such collection of proofs without words (PWWs). I should note that this collection, like the first two, is necessarily incomplete. It does not include all PWWs that have appeared in print since the second collection appeared, or all of those that I overlooked in compiling the first two books. As readers of the Association's journals are well aware, new PWWs appear in print rather frequently, and they also appear now on the World Wide Web in formats superior to print, involving motion and viewer interaction.

I hope that the readers of this collection will find enjoyment in discovering or rediscovering some elegant visual demonstrations of certain mathematical ideas, that teachers will share them with their students, and that all will find stimulation and encouragement to create new proofs without words.

Acknowledgment. I would like to express my appreciation and gratitude to all those individuals who have contributed proofs without words to the mathematical literature; see the *Index of Names* on pp. 185–186. Without them this collection simply would not exist. Thanks to Susan Staples and the members of the editorial board of Classroom Resource Materials for their careful reading of an earlier draft of this book, and for their many helpful suggestions. I would also like to thank Carol Baxter, Beverly Ruedi, and Samantha Webb of the MAA's publication staff for their encouragement, expertise, and hard work in preparing this book for publication.

Roger B. Nelsen
Lewis & Clark College
Portland, Oregon

Notes

1. The illustrations in this collection were redrawn to create a uniform appearance. In a few instances titles were changed, and shading or symbols were added or deleted for clarity. Any errors resulting from that process are entirely my responsibility.

2. Roman numerals are used in the titles of some PWWs to distinguish multiple PWWs of the same theorem—and the numbering is carried over from *Proofs Without Words* and *Proofs Without Words II*. So, for example, since there are six PWWs of the Pythagorean Theorem in *Proofs Without Words* and six more in *Proofs Without Words II*, the first in this collection carries the title "The Pythagorean Theorem XIII."

3. Several PWWs in this collection are presented in the form of "solutions" to problems from mathematics contests such as the William Lowell Putnam Mathematical Competition and the Kazakh National Mathematical Olympiad. It is quite doubtful that such "solutions" would have garnered many points in those contests, as contestants are advised in, for example, the Putnam competition that "all the necessary steps of a proof must be shown clearly to obtain full credit."

Contents

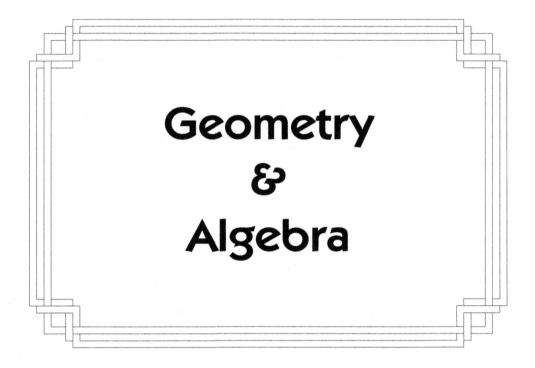

Geometry
&
Algebra

1

The Pythagorean Theorem XIII

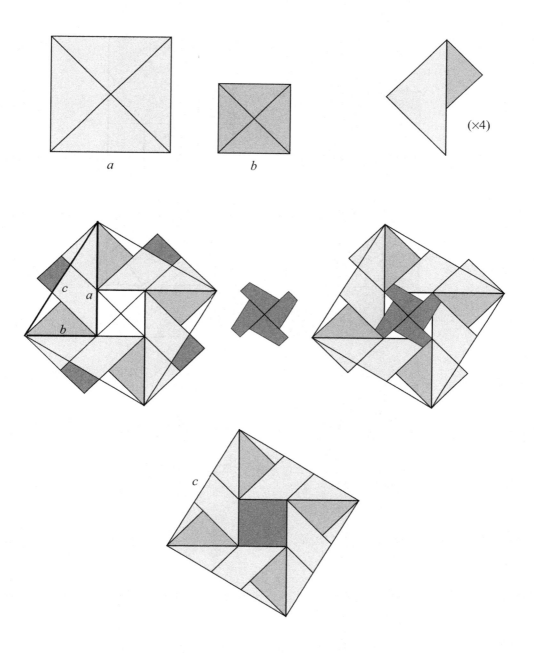

—José A. Gomez

The Pythagorean Theorem XIV

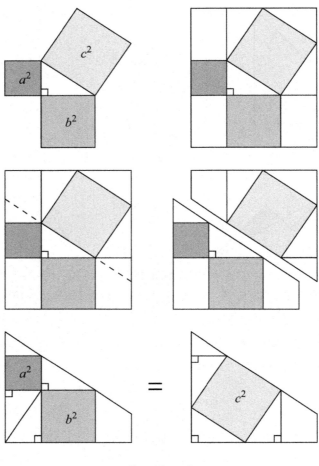

$$a^2 + b^2 = c^2.$$

The Pythagorean Theorem XV

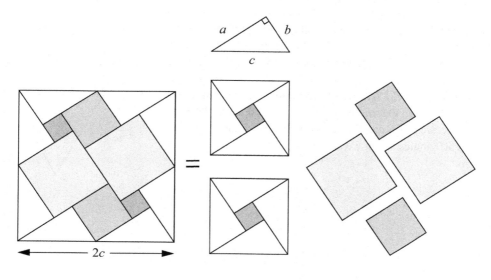

$$(2c)^2 = 2c^2 + 2a^2 + 2b^2$$
$$\therefore c^2 = a^2 + b^2.$$

—Nam Gu Heo

The Pythagorean Theorem XVI

The Pythagorean theorem (Proposition I.47 in Euclid's *Elements*) is usually illustrated with squares drawn on the sides of a right triangle. However, as a consequence of Proposition VI.31 in the *Elements*, any set of three similar figures may be used, such as equilateral triangles as shown at the right. Let T denote the area of a right triangle with legs a and b and hypotenuse c, let T_a, T_b, and T_c denote the areas of equilateral triangles drawn externally on sides a, b, and c, and let P denote the area of a parallelogram with sides a and b and $30°$ and $150°$ angles. Then we have

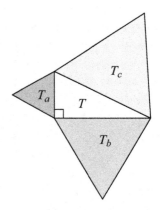

1. $T = P$.

Proof.

2. $T_c = T_a + T_b$.

Proof.

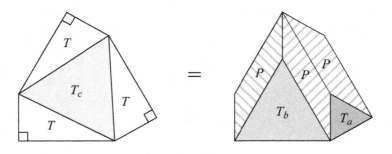

—Claudi Alsina & RBN

Pappus' Generalization of the Pythagorean Theorem

Let ABC be any triangle, and $ABDE$, $ACFG$ any parallelograms described externally on AB and AC. Extend DE and FG to meet in H and draw BL and CM equal and parallel to HA. Then, in area, $BCML = ABDE + ACFG$ [*Mathematical Collection*, Book IV].

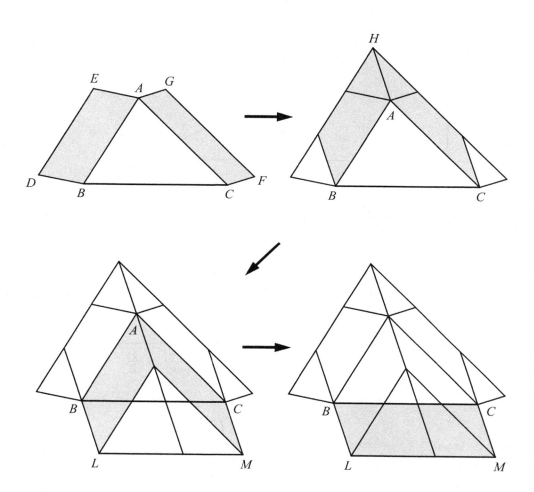

—Pappus of Alexandria (circa 320 CE)

A Reciprocal Pythagorean Theorem

If a and b are the legs and h the altitude to the hypotenuse c of a right triangle, then

$$\frac{1}{a^2} + \frac{1}{b^2} = \frac{1}{h^2}.$$

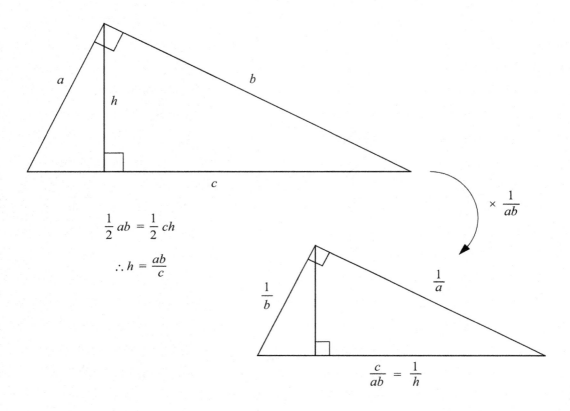

$$\frac{1}{2}\,ab = \frac{1}{2}\,ch$$

$$\therefore h = \frac{ab}{c}$$

$$\times \frac{1}{ab}$$

$$\frac{c}{ab} = \frac{1}{h}$$

$$\therefore \left(\frac{1}{a}\right)^2 + \left(\frac{1}{b}\right)^2 = \left(\frac{1}{h}\right)^2.$$

NOTE: For another proof, see Vincent Ferlini, Mathematics without (many) words, *College Mathematics Journal* **33** (2002), p. 170.

—RBN

A Pythagorean-Like Formula

Given an isosceles triangle as shown in the figure, we have

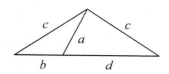

$$c^2 = a^2 + bd.$$

Proof.

1.

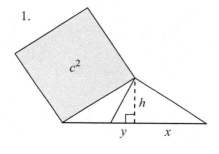

$$x + y = d$$
$$x - y = b$$

2.

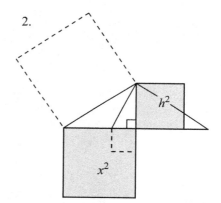

3.

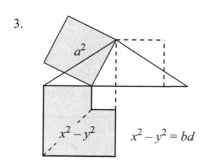

$$x^2 - y^2 = bd$$

4.

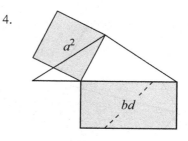

—Larry Hoehn

Four Pythagorean-Like Theorems

Let T denote the area of a triangle with angles α, β, and γ; and let T_α, T_β, and T_γ denote the areas of equilateral triangles constructed externally on the sides opposite α, β, and γ. Then the following theorems hold:

I. *If $\alpha = \pi/3$, then $T + T_\alpha = T_\beta + T_\gamma$.*

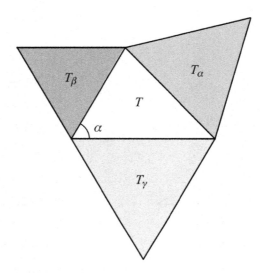

Proof.

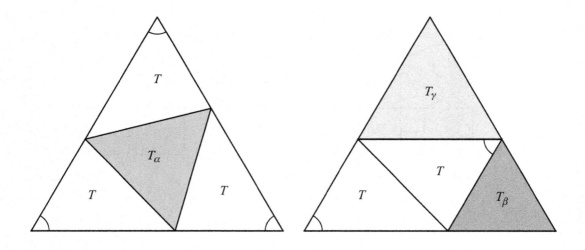

—Manuel Moran Cabre

II. *If* $\alpha = 2\pi/3$, *then* $T_\alpha = T_\beta + T_\gamma + T$.

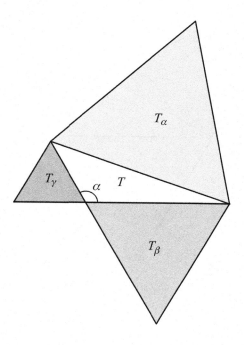

Proof.

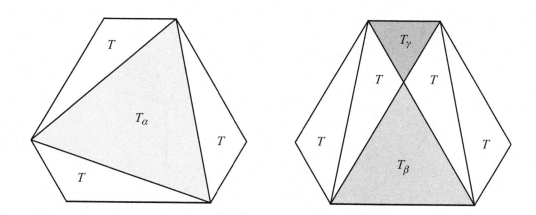

—RBN

III. *If $\alpha = \pi/6$, then $T_\alpha + 3T = T_\beta + T_\gamma$.*

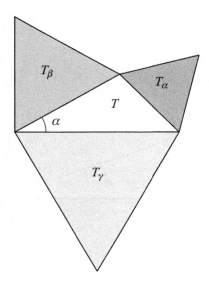

Proof.

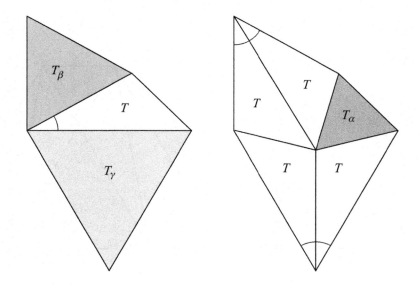

—Claudi Alsina & RBN

IV. *If $\alpha = 5\pi/6$, then $T_\alpha = T_\beta + T_\gamma + 3T$.*

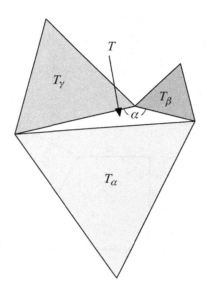

Proof.

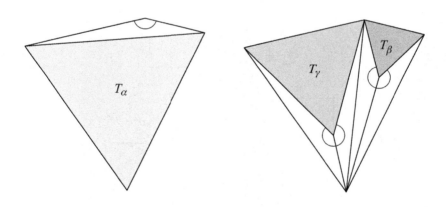

NOTE: In general, $T_\alpha = T_\beta + T_\gamma - \sqrt{3}\,T\cot\alpha$.

—Claudi Alsina & RBN

Pythagoras for a Right Trapezoid

A *right trapezoid* is a trapezoid with two right angles. If a and b are the lengths of the bases, h the height, s the slant height, and c and d the diagonals, then

$$c^2 + d^2 = s^2 + h^2 + 2ab.$$

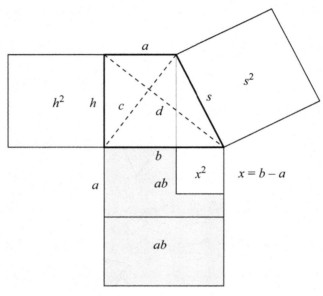

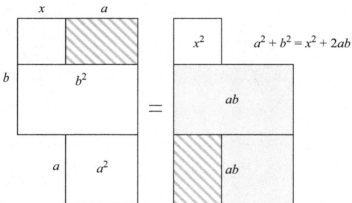

$$c^2 + d^2 = (a^2 + h^2) + (b^2 + h^2) = x^2 + 2ab + 2h^2 = s^2 + h^2 + 2ab.$$

—Guanshen Ren

Pythagoras for a Clipped Rectangle

An (a, b, c)-*clipped rectangle* is an $a \times b$ rectangle where one corner has been cut off to form a fifth side of length c. If d and e are the lengths of the two diagonals nearest the fifth side, then

$$a^2 + b^2 + c^2 = d^2 + e^2.$$

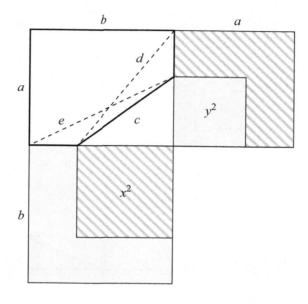

$$a^2 + b^2 + c^2 = a^2 + b^2 + (x^2 + y^2) = (a^2 + x^2) + (b^2 + y^2) = d^2 + e^2.$$

—Guanshen Ren

Heron's Formula

(Heron of Alexandria, circa 10–70 CE)

The area K of a triangle with sides a, b, and c and semiperimeter $s = (a+b+c)/2$ is $K = \sqrt{s(s-a)(s-b)(s-c)}$.

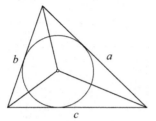 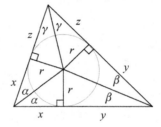

$$s = x+y+z, \, x = s-a, \, y = s-b, \, z = s-c.$$

1. $K = r(x+y+z) = rs.$

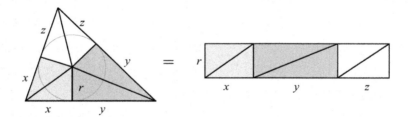

2. $xyz = r^2(x+y+z) = r^2s.$

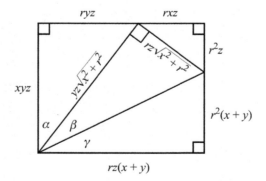

3. $\therefore K^2 = r^2s^2 = sxyz = s(s-a)(s-b)(s-c).$

—RBN

Every Triangle Has Infinitely Many Inscribed Equilateral Triangles

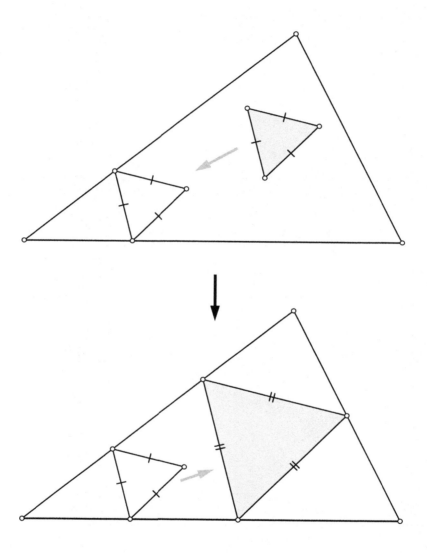

—Sidney H. Kung

Every Triangle Can Be Dissected into Six Isosceles Triangles

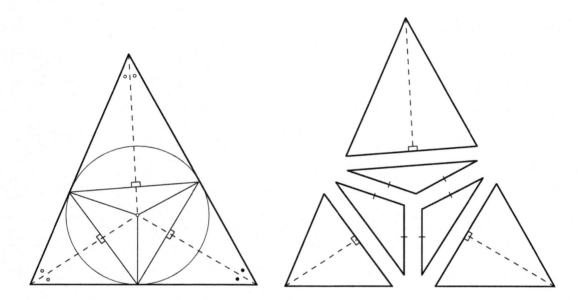

—Ángel Plaza

More Isosceles Dissections

1. Every triangle can be dissected into four isosceles triangles:

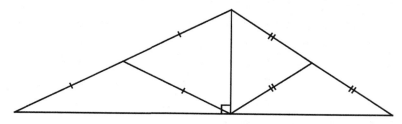

2. Every acute triangle can be dissected into three isosceles triangles:

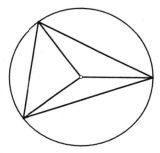

3. A triangle can be dissected into two isosceles triangles if it is a right triangle or if one of its angles is two or three times another:

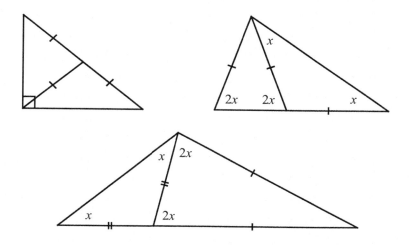

—Des MacHale

Viviani's Theorem II

(Vincenzo Viviani, 1622–1703)

In an equilateral triangle, the sum of the distances from any interior point to the three sides equals the altitude of the triangle.

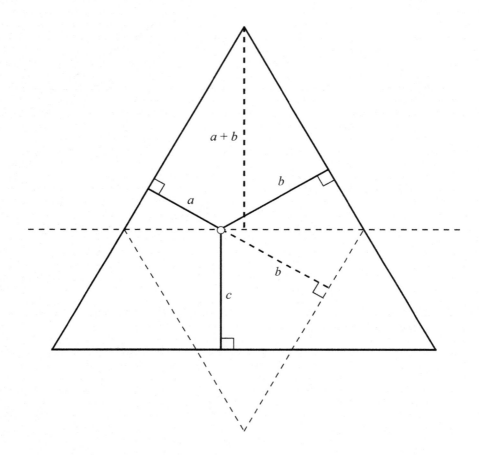

—James Tanton

Viviani's Theorem III

In an equilateral triangle, the sum of the distances from an interior point to the three sides equals the altitude of the triangle.

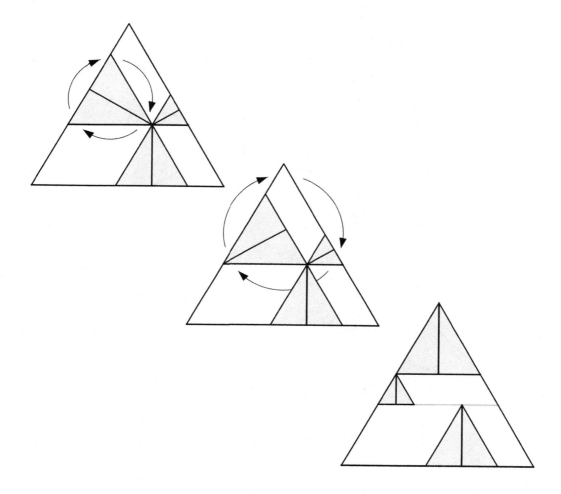

—Ken-ichiroh Kawasaki

Ptolemy's Theorem I

In a quadrilateral inscribed in a circle, the product of the lengths of the diagonals is equal to the sum of the products of the lengths of the opposite sides.

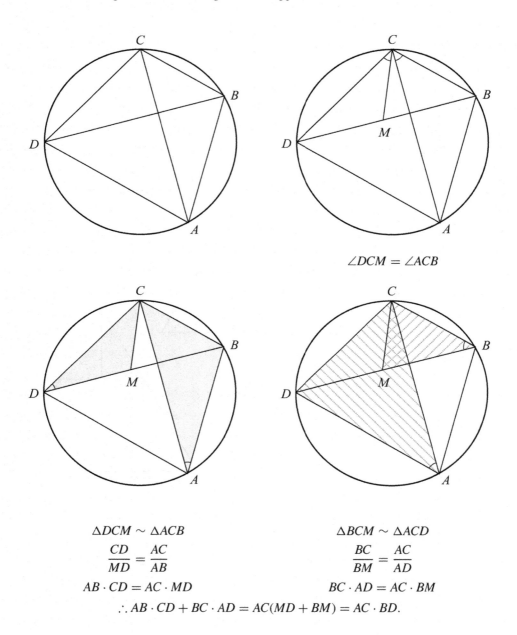

$$\angle DCM = \angle ACB$$

$$\triangle DCM \sim \triangle ACB \qquad\qquad \triangle BCM \sim \triangle ACD$$

$$\frac{CD}{MD} = \frac{AC}{AB} \qquad\qquad\qquad \frac{BC}{BM} = \frac{AC}{AD}$$

$$AB \cdot CD = AC \cdot MD \qquad\qquad BC \cdot AD = AC \cdot BM$$

$$\therefore AB \cdot CD + BC \cdot AD = AC(MD + BM) = AC \cdot BD.$$

—Ptolemy of Alexandria (circa 90–168 CE)

Ptolemy's Theorem II

In a quadrilateral inscribed in a circle, the product of the lengths of the diagonals is equal to the sum of the products of the lengths of the opposite sides.

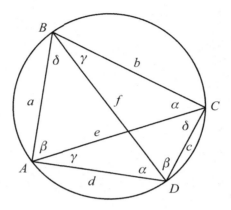

$$\alpha + \beta + \gamma + \delta = 180°$$

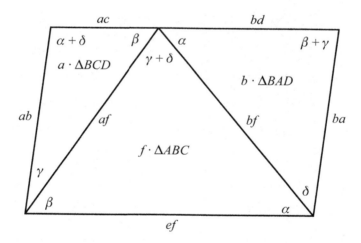

$$\therefore ef = ac + bd.$$

—William Derrick & James Hirstein

Equal Areas in a Partition of a Parallelogram

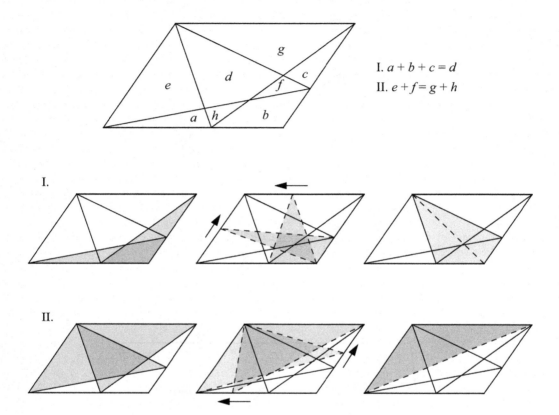

I. $a + b + c = d$

II. $e + f = g + h$

—Philippe R. Richard

The Area of an Inner Square

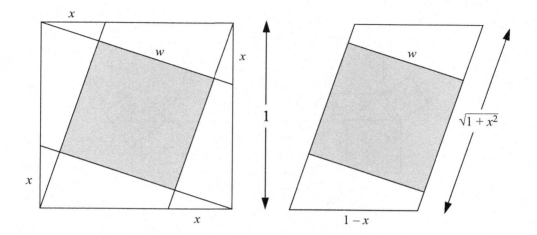

$$\text{Area } \varnothing = w \cdot \sqrt{1 + x^2} = 1 \cdot (1 - x),$$

$$\text{Area } \blacksquare = w^2 = \frac{(1 - x)^2}{1 + x^2}.$$

—Marc Chamberland

The Parallelogram Law

In any parallelogram, the sum of the squares of the sides is equal to the sum of the squares of the diagonals.

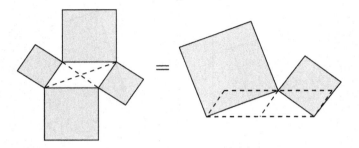

Proof.

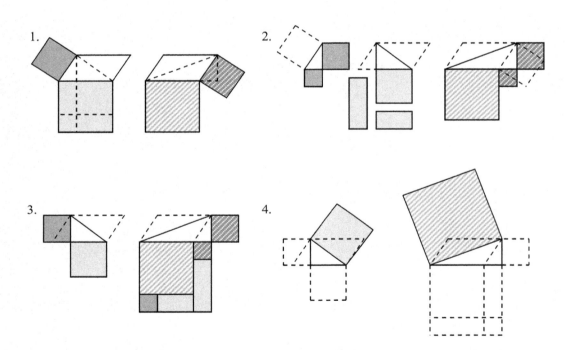

—Claudi Alsina & Amadeo Monreal

The Length of a Triangle Median via the Parallelogram Law

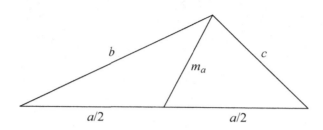

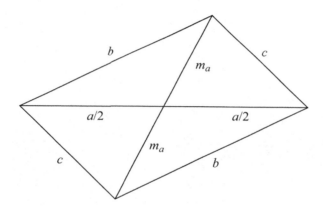

$$2b^2 + 2c^2 = a^2 + (2m_a)^2,$$

$$\therefore m_a = \frac{1}{2}\sqrt{2b^2 + 2c^2 - a^2}.$$

—C. Peter Lawes

Two Squares and Two Triangles

If two squares share a corner, then the vertical triangles on either side of that point have equal area.

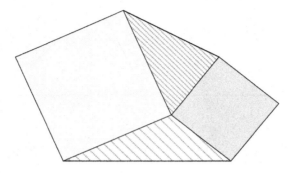

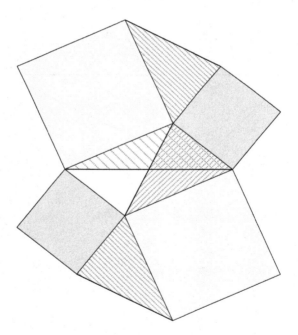

The Inradius of an Equilateral Triangle

The inradius of an equilateral triangle is one-third the height of the triangle.

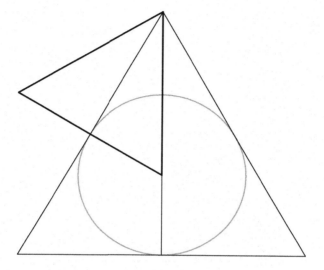

—Participants of the Summer Institute Series
2004 Geometry Course
School of Education, Northeastern University
Boston, MA 02115

A Line Through the Incenter of a Triangle

A line passing through the incenter of a triangle bisects the perimeter if and only if it bisects the area.

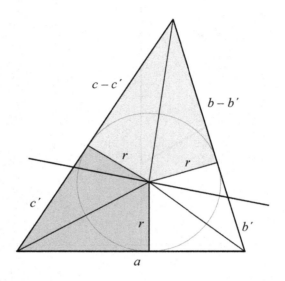

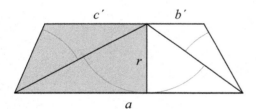

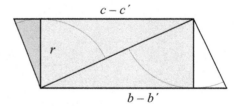

$$A_{\text{bottom}} = A_{\text{top}} \quad \Leftrightarrow \quad a + b' + c' = c - c' + b - b' = \frac{a+b+c}{2}.$$

—Sidney H. Kung

The Area and Circumradius of a Triangle

If K, a, b, c, and R denote, respectively, the area, lengths of the sides, and circumradius of a triangle, then

$$K = \frac{abc}{4R}.$$

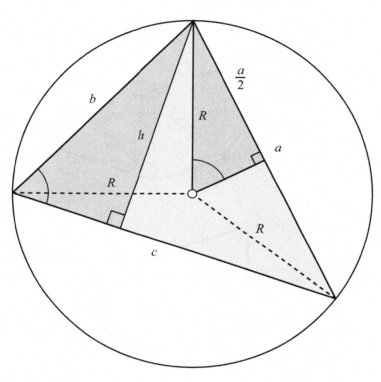

$$\frac{h}{b} = \frac{a/2}{R} \quad \Rightarrow \quad h = \frac{1}{2}\frac{ab}{R},$$
$$\therefore K = \frac{1}{2}hc = \frac{1}{4}\frac{abc}{R}.$$

Beyond Extriangles

For any $\triangle ABC$, construct squares on each of the three sides. Connecting adjacent square corners creates three extriangles. Iterating this process produces three quadrilaterals, each with area five times the area of $\triangle ABC$. In the figure, letting [ ] denote area, we have

$$[A_1A_2A_3A_4] = [B_1B_2B_3B_4] = [C_1C_2C_3C_4] = 5[ABC].$$

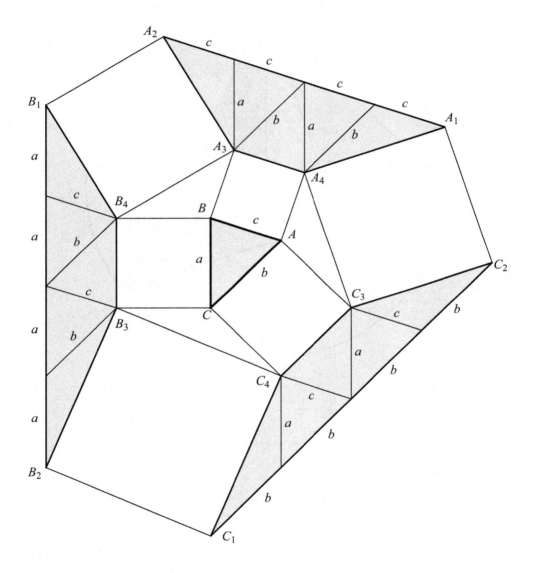

—M. N. Deshpande

A 45° Angle Sum

(Problem 3, Student Mathematics Competition of the Illinois Section of the MAA, 2001)

Suppose $ABCD$ is a square and n is a positive integer. Let $X_1, X_2, X_3, \cdots, X_n$ be points on BC so that $BX_1 = X_1X_2 = \cdots = X_{n-1}X_n = X_nC$. Let Y be the point on AD so that $AY = BX_1$. Find (in degrees) the value of

$$\angle AX_1Y + \angle AX_2Y + \cdots + \angle AX_nY + \angle ACY.$$

Solution. The value of the sum is $45°$. Proof (for $n = 4$):

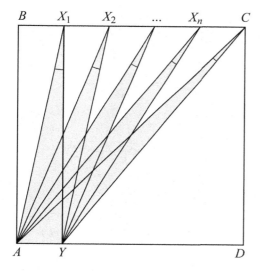 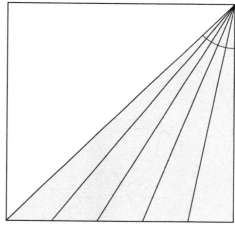

Trisection of a Line Segment II

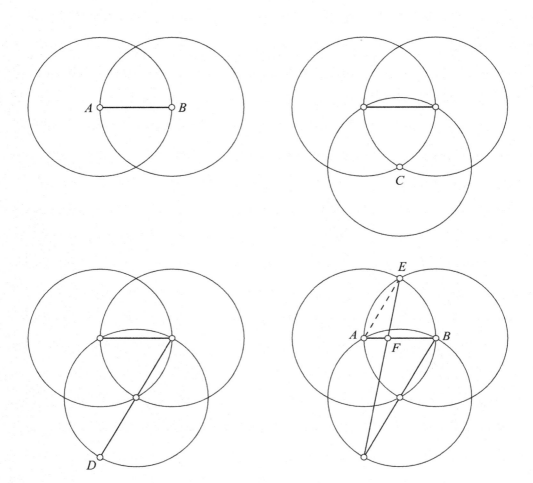

$$\overline{AF} = \frac{1}{3} \cdot \overline{AB}$$

—Robert Styer

Two Squares with Constant Area

If a diameter of a circle intersects a chord of the circle at 45°, cutting off segments of the chord of lengths a and b, then $a^2 + b^2$ is constant.

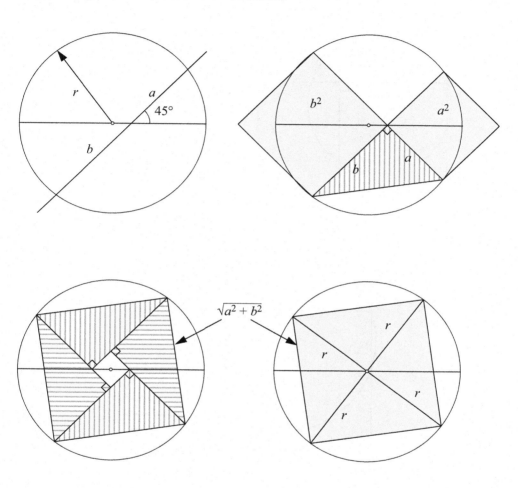

$$a^2 + b^2 = 2r^2.$$

Four Squares with Constant Area

If two chords of a circle intersect at right angles, then the sum of the squares of the lengths of the four segments formed is constant (and equal to the square of the length of the diameter).

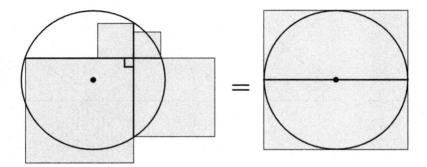

Proof.

1.
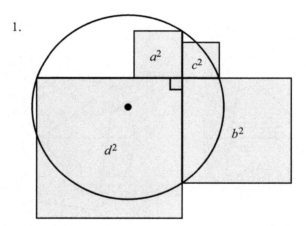

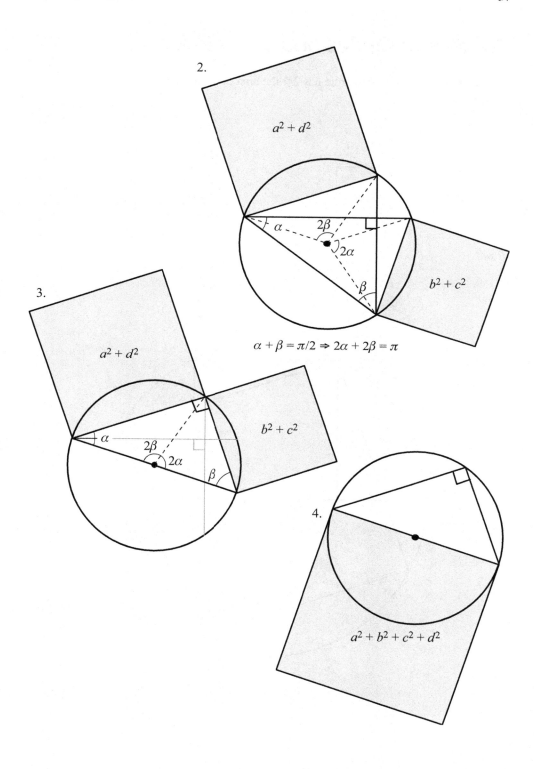

2.

$a^2 + d^2$

α 2β 2α β

$b^2 + c^2$

$\alpha + \beta = \pi/2 \Rightarrow 2\alpha + 2\beta = \pi$

3.

$a^2 + d^2$

α 2β 2α β

$b^2 + c^2$

4.

$a^2 + b^2 + c^2 + d^2$

—RBN

Squares in Circles and Semicircles

A square inscribed in a semicircle has 2/5 the area of a square inscribed in a circle of the same radius.

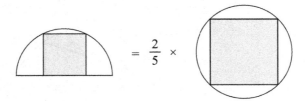

Proof.

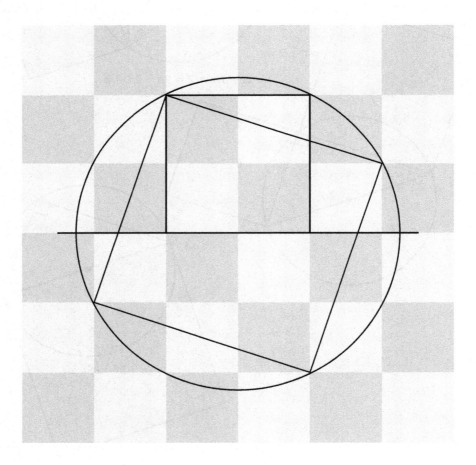

—RBN

The Christmas Tree Problem

(Problem 370, *Journal of Recreational Mathematics*, **8** (1976), p. 46)

An isosceles right triangle is inscribed in a semicircle, and
the radius bisecting the other semicircle is drawn. Circles
are inscribed in the triangle and the two quadrants as shown.
Prove that these three smaller circles are congruent.

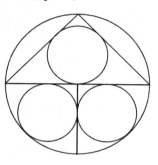

Solution.

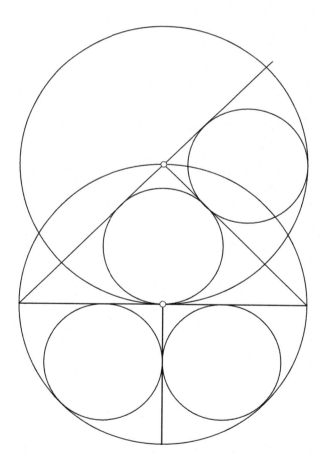

The Area of an Arbelos

Theorem. Let P, Q, and R be three points on a line, with Q lying between P and R. Semicircles are drawn on the same side of the line with diameters PQ, QR, and PR. An *arbelos* is the figure bounded by these three semicircles. Draw the perpendicular to PR at Q, meeting the largest semicircle at S. Then the area A of the arbelos equals the area C of the circle with diameter QS [Archimedes, *Liber Assumptorum*, Proposition 4].

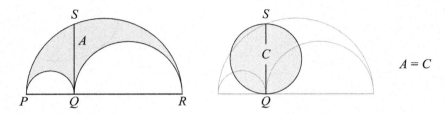

Proof.

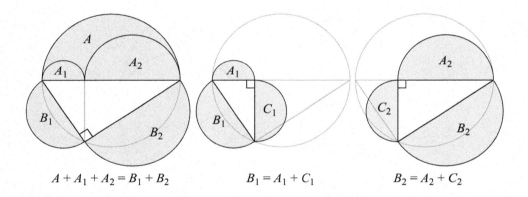

$$A + A_1 + A_2 = B_1 + B_2 \qquad B_1 = A_1 + C_1 \qquad B_2 = A_2 + C_2$$

$$A + A_1 + A_2 = A_1 + C_1 + A_2 + C_2$$
$$\therefore A = C_1 + C_2 = C$$

—RBN

The Area of a Salinon

Theorem. Let P, Q, R, S be four points on a line (in that order) such that $PQ = RS$. Semicircles are drawn above the line with diameters PQ, RS, and PS, and another semicircle with diameter QR is drawn below the line. A *salinon* is the figure bounded by these four semicircles. Let the axis of symmetry of the salinon intersect its boundary at M and N. Then the area A of the salinon equals the area C of the circle with diameter MN [Archimedes, *Liber Assumptorum*, Proposition 14].

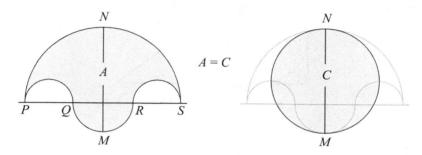

Proof.

1.

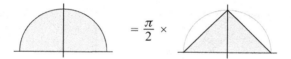

2.

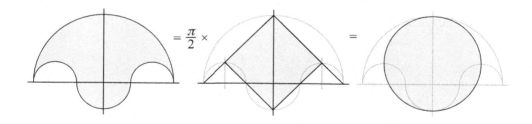

—RBN

The Area of a Right Triangle

Theorem. The area K of a right triangle is equal to the product of the lengths of the segments of the hypotenuse determined by the point of tangency of the inscribed circle.

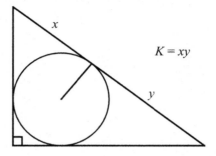

$$K = xy$$

Proof.

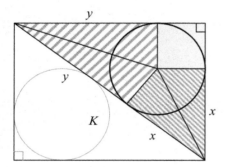

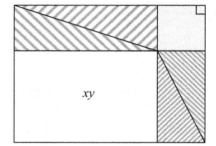

—RBN

The Area of a Regular Dodecagon II

A regular dodecagon inscribed in a circle of radius one has area three.

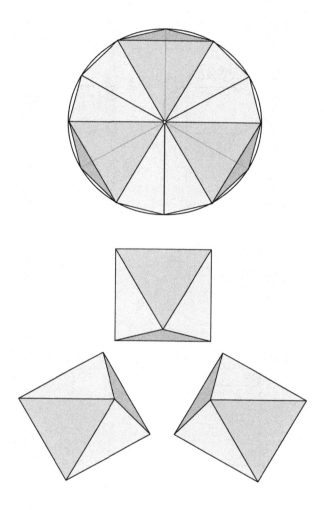

—RBN

Four Lunes Equal One Square

Theorem. If a square is inscribed in a circle and four semicircles constructed on its sides, then the area of the four lunes equals the area of the square [Hippocrates of Chios, circa 440 BCE].

Proof.

Lunes and the Regular Hexagon

Theorem. If a regular hexagon is inscribed in a circle and six semicircles constructed on its sides, then the area of the hexagon equals the area of the six lunes plus the area of a circle whose diameter is equal in length to one of the sides of the hexagon [Hippocrates of Chios, circa 440 BCE].

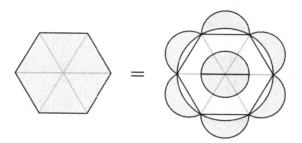

Proof.

$$[4 \cdot \pi (r/2)^2 = \pi r^2]$$

—RBN

The Volume of a Triangular Pyramid

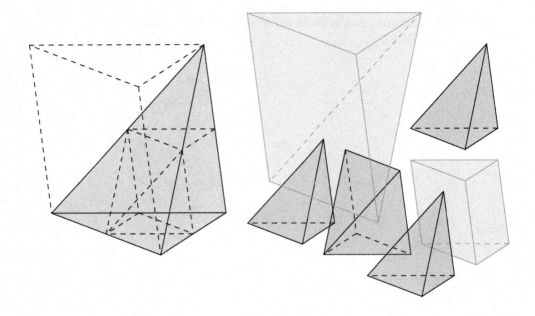

$$V_{\text{Prism}} = \left(V_{\text{Prism}} - V_{\text{Pyramid}}\right) + 3 \times \frac{1}{8}V_{\text{Pyramid}}$$

$$+ \frac{1}{8}V_{\text{Prism}} + \frac{1}{8}\left(V_{\text{Prism}} - V_{\text{Pyramid}}\right)$$

$$\therefore V_{\text{Pyramid}} = \frac{1}{3}V_{\text{Prism}}$$

—Poo-Sung Park

Algebraic Areas IV

I. $ax - by = \dfrac{1}{2}(a+b)(x-y) + \dfrac{1}{2}(a-b)(x+y)$

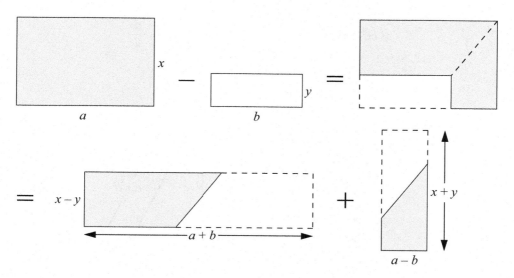

II. $ax + by = \dfrac{1}{2}(a+b)(x+y) + \dfrac{1}{2}(a-b)(x-y)$

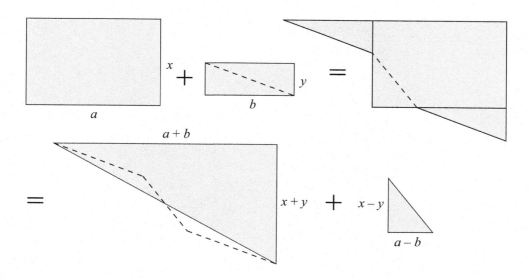

—Yukio Kobayashi

Componendo et Dividendo, a Theorem on Proportions

If $bd \neq 0$ and $\dfrac{a}{b} = \dfrac{c}{d} \neq 1$, then $\dfrac{a+b}{a-b} = \dfrac{c+d}{c-d}$.

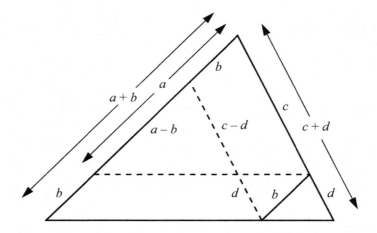

—Yukio Kobayashi

Completing the Square II

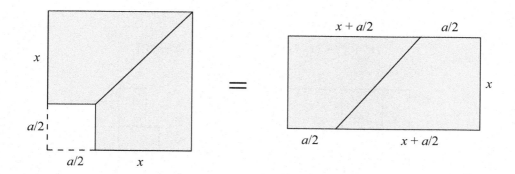

$$\left(x + \frac{a}{2}\right)^2 - \left(\frac{a}{2}\right)^2 = x(x+a) = x^2 + ax.$$

—Munir Mahmood

Candido's Identity

(Giacomo Candido, 1871–1941)

$$[x^2 + y^2 + (x+y)^2]^2 = 2[x^4 + y^4 + (x+y)^4]$$

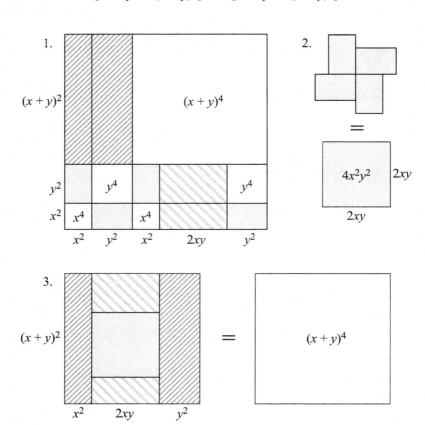

NOTE: Candido employed this identity to establish $[F_n^2 + F_{n+1}^2 + F_{n+2}^2]^2 = 2[F_n^4 + F_{n+1}^4 + F_{n+2}^4]$, where F_n denotes the nth Fibonacci number.

—RBN

Trigonometry,
Calculus,
&
Analytic Geometry

Sine of a Sum or Difference (via the Law of Sines)

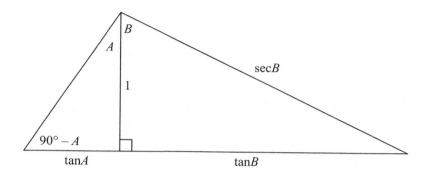

$$\frac{\sin(A + B)}{\tan A + \tan B} = \frac{\sin(90° - A)}{\sec B}$$

$$\therefore \sin(A + B) = \cos A \cos B(\tan A + \tan B)$$
$$= \sin A \cos B + \cos A \sin B$$

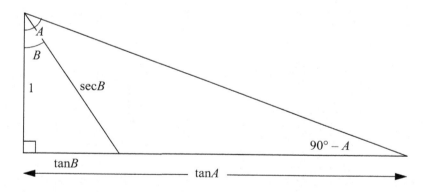

$$\frac{\sin(A - B)}{\tan A - \tan B} = \frac{\sin(90° - A)}{\sec B}$$

$$\therefore \sin(A - B) = \cos A \cos B(\tan A - \tan B)$$
$$= \sin A \cos B - \cos A \sin B$$

—James Kirby

Cosine of the Difference I

$$\cos(\alpha - \beta) = \cos\alpha\cos\beta + \sin\alpha\sin\beta.$$

I.

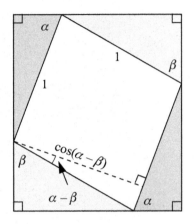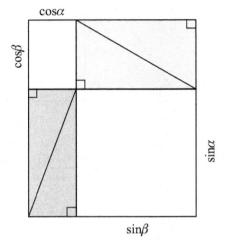

II.

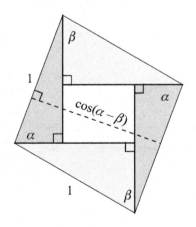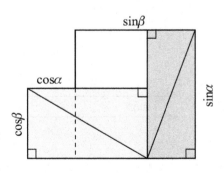

—William T. Webber & Matthew Bode

Sine of the Sum IV and Cosine of the Difference II

I. $\sin(u + v) = \sin u \cos v + \sin v \cos u$.

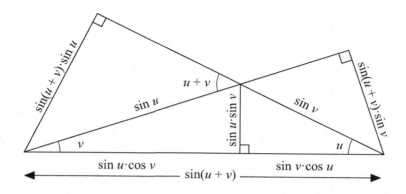

—Long Wang

II. $\cos(u - v) = \cos u \cos v + \sin u \sin v$.

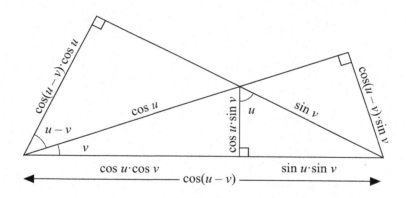

—David Richeson

The Double Angle Formulas IV

$$\sin 2x = 2 \sin x \cos x \text{ and } \cos 2x = \cos^2 x - \sin^2 x$$

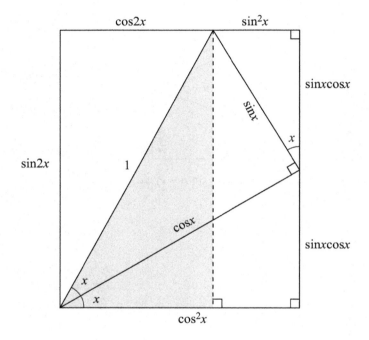

—Hasan Unal

Euler's Half Angle Tangent Formula

(Leonhard Euler, 1707–1783)

$$\tan \frac{\alpha + \beta}{2} = \frac{\sin \alpha + \sin \beta}{\cos \alpha + \cos \beta}$$

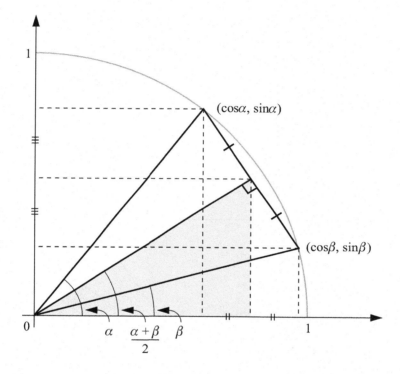

$$\tan \frac{\alpha + \beta}{2} = \frac{(\sin \alpha + \sin \beta)/2}{(\cos \alpha + \cos \beta)/2}.$$

—Don Goldberg

The Triple Angle Sine and Cosine Formulas I

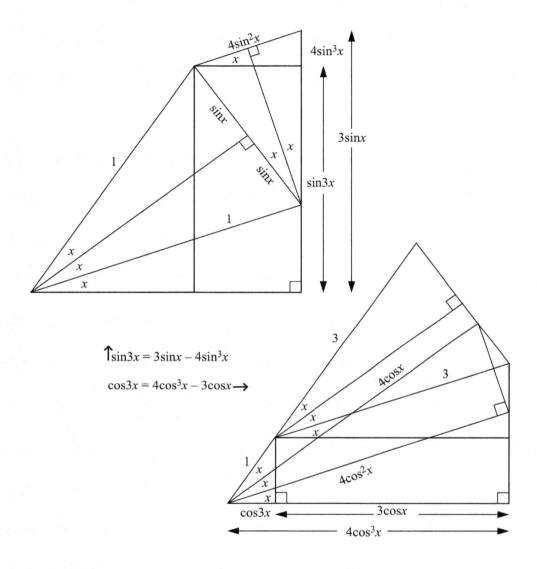

$$\uparrow \sin 3x = 3\sin x - 4\sin^3 x$$

$$\cos 3x = 4\cos^3 x - 3\cos x \rightarrow$$

—Shingo Okuda

The Triple Angle Sine and Cosine Formulas II

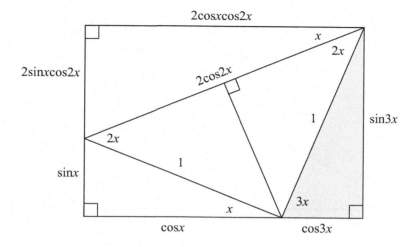

$$\sin 3x = 2 \sin x \cos 2x + \sin x,$$
$$= 2 \sin x (1 - 2 \sin^2 x) + \sin x,$$
$$= 3 \sin x - 4 \sin^3 x;$$
$$\cos 3x = 2 \cos x \cos 2x - \cos x,$$
$$= 2 \cos x (2 \cos^2 x - 1) - \cos x,$$
$$= 4 \cos^3 x - 3 \cos x.$$

—Claudi Alsina & RBN

Trigonometric Functions of 15° and 75°

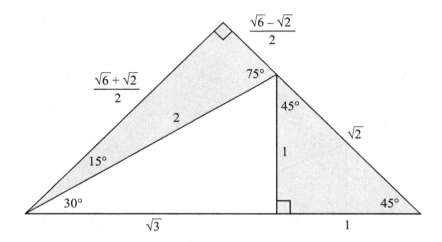

$$\sin 15° = \frac{\sqrt{6} - \sqrt{2}}{4}, \quad \tan 75° = \frac{\sqrt{6} + \sqrt{2}}{\sqrt{6} - \sqrt{2}}, \quad \text{etc.}$$

Corollary. Areas of shaded triangles are equal (to $1/2$).

—Larry Hoehn

Trigonometric Functions of Multiples of 18°

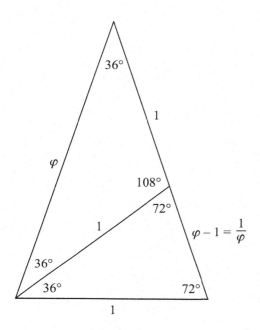

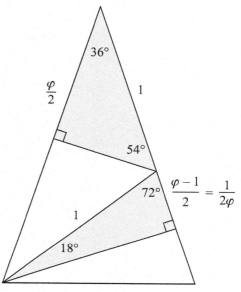

$$\frac{\varphi}{1} = \frac{1}{\varphi - 1}$$

$$\varphi^2 - \varphi - 1 = 0$$

$$\varphi = \frac{\sqrt{5} + 1}{2}$$

$$\sin 54° = \cos 36° = \frac{\varphi}{2} = \frac{\sqrt{5} + 1}{4}$$

$$\sin 18° = \cos 72° = \frac{1}{2\varphi} = \frac{1}{\sqrt{5} + 1}$$

—Brian Bradie

Mollweide's Equation II

(Karl Brandan Mollweide, 1774–1825)

$$\frac{\sin\left((\alpha - \beta)/2\right)}{\cos\left(\gamma/2\right)} = \frac{a - b}{c}$$

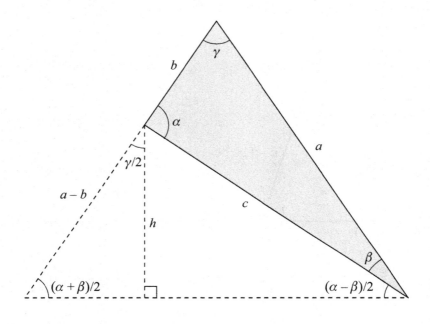

$$\frac{\sin\left((\alpha - \beta)/2\right)}{\cos\left(\gamma/2\right)} = \frac{h/c}{h/(a - b)} = \frac{a - b}{c}.$$

NOTE: For another proof of this identity by the same author, see Mathematics without words II, *College Mathematics Journal* **32** (2001), 68–69.

—Rex H. Wu

Newton's Formula (for the General Triangle)

(Sir Isaac Newton, 1642–1726)

$$\frac{\cos((\alpha - \beta)/2)}{\sin(\gamma/2)} = \frac{a + b}{c}$$

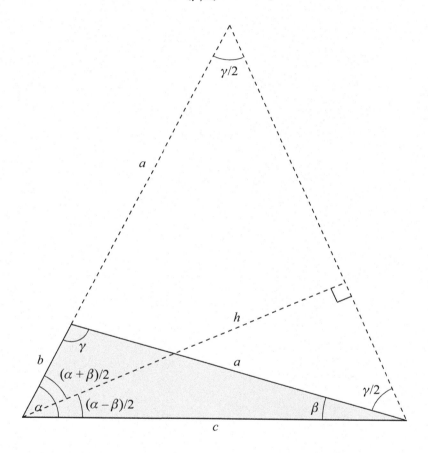

$$\frac{\cos((\alpha - \beta)/2)}{\sin(\gamma/2)} = \frac{h/c}{h/(a+b)} = \frac{a+b}{c}.$$

A Sine Identity for Triangles

$$x + y + z = \pi \quad \Rightarrow \quad 4 \sin x \sin y \sin z = \sin 2x + \sin 2y + \sin 2z$$

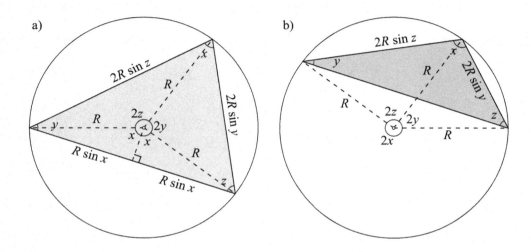

a) $\quad \dfrac{1}{2}(2R \sin y)(2R \sin z) \sin x = \dfrac{1}{2}R^2 \sin 2x + \dfrac{1}{2}R^2 \sin 2y + \dfrac{1}{2}R^2 \sin 2z.$

b) $\quad \dfrac{1}{2}(2R \sin y)(2R \sin z) \sin x = \dfrac{1}{2}R^2 \sin 2y + \dfrac{1}{2}R^2 \sin 2z - \dfrac{1}{2}R^2 \sin(2\pi - 2x).$

NOTE: The identity actually holds for all real x, y, z such that $x + y + z = \pi$.

—RBN

Cofunction Sums

$$\sin x + \cos x = \sqrt{2}\sin\left(x + \frac{\pi}{4}\right) \qquad \tan x + \cot x = 2\csc(2x)$$

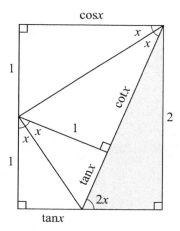

$$\csc x + \cot x = \cot(x/2)$$

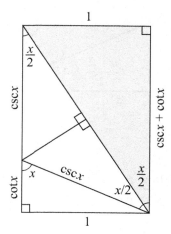

Corollary. $\cos x - \sin x = \sqrt{2}\cos(x + \pi/4)$ and $\cot x - \tan x = 2\cot(2x)$.

—RBN

The Law of Tangents I

$$\frac{\tan((\alpha - \beta)/2)}{\tan((\alpha + \beta)/2)} = \frac{a - b}{a + b}$$

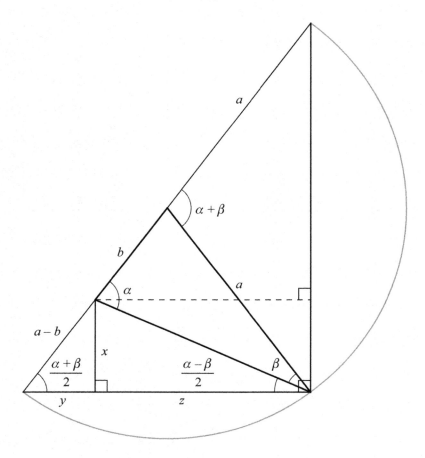

$$\frac{\tan((\alpha - \beta)/2)}{\tan((\alpha + \beta)/2)} = \frac{x/z}{x/y} = \frac{y}{z} = \frac{a - b}{a + b}.$$

—Rex H. Wu

The Law of Tangents II

$$\frac{\tan((\alpha - \beta)/2)}{\tan((\alpha + \beta)/2)} = \frac{a - b}{a + b}$$

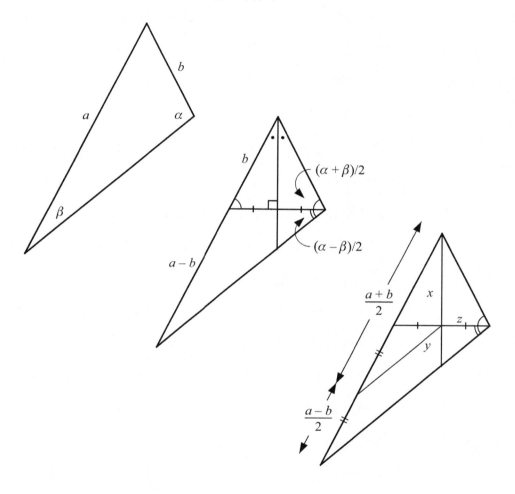

$$\frac{\tan((\alpha - \beta)/2)}{\tan((\alpha + \beta)/2)} = \frac{y/z}{x/z} = \frac{(a-b)/2}{(a+b)/2} = \frac{a - b}{a + b}.$$

—Wm. F. Cheney, Jr.

Need a Solution to $x + y = xy$?

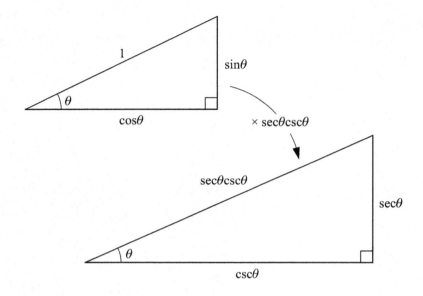

$$\sec^2 \theta + \csc^2 \theta = \sec^2 \theta \csc^2 \theta.$$

—RBN

An Identity for $\sec x + \tan x$

$$\sec x + \tan x = \tan\left(\frac{\pi}{4} + \frac{x}{2}\right)$$

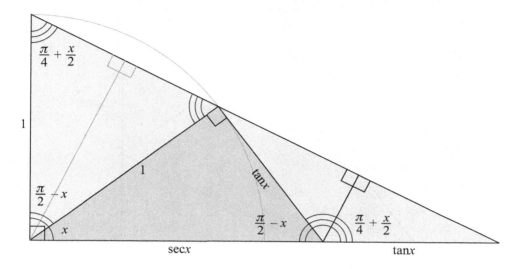

NOTE: Calculus students will recognize the expression $\sec x + \tan x$ since it appears in the indefinite integral of the secant of x. However, the first known formula for this integral, discovered in 1645, was

$$\int \sec x \, dx = \ln\left|\tan\left(\frac{\pi}{4} + \frac{x}{2}\right)\right| + C.$$

For details see V. F. Rickey and P. M. Tuchinsky, "An Application of Geography to Mathematics: History of the Integral of the Secant," *Mathematics Magazine*, **53** (1980), pp. 162–166.

—RBN

A Sum of Tangent Products

If α, β, and γ are any positive angles such that $\alpha + \beta + \gamma = \pi/2$, then

$$\tan\alpha\tan\beta + \tan\beta\tan\gamma + \tan\gamma\tan\alpha = 1.$$

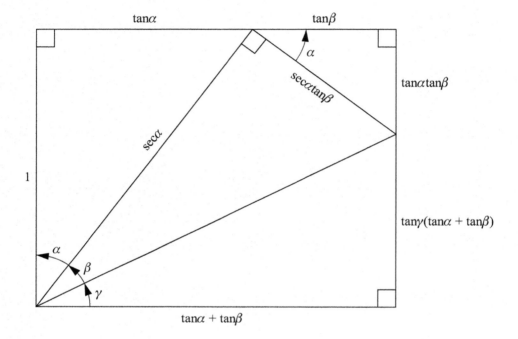

—RBN

A Sum and Product of Three Tangents

If α, β, and γ denote angles in an acute triangle, then

$$\tan\alpha + \tan\beta + \tan\gamma = \tan\alpha\,\tan\beta\,\tan\gamma.$$

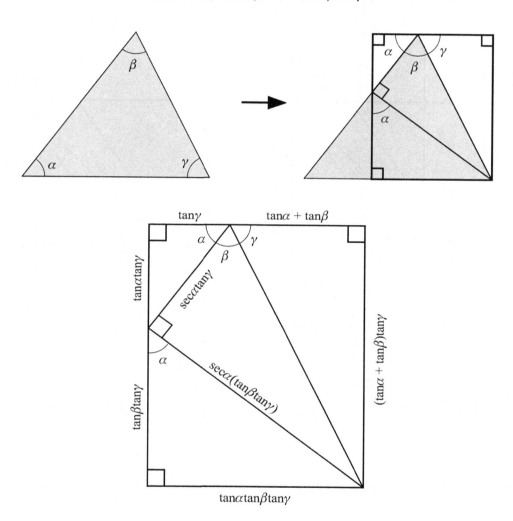

NOTE: The result holds for any angles α, β, γ (none an odd multiple of $\pi/2$) whose sum is π.

—RBN

A Product of Tangents

$$\tan(\pi/4 + \alpha) \cdot \tan(\pi/4 - \alpha) = 1$$

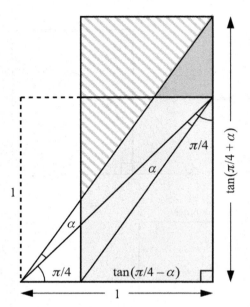 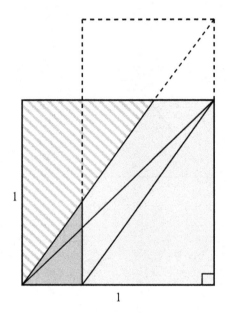

—RBN

Sums of Arctangents II

$$0 < m < n \quad \Rightarrow \quad \arctan\left(\frac{m}{n}\right) + \arctan\left(\frac{n-m}{n+m}\right) = \frac{\pi}{4}$$

I.

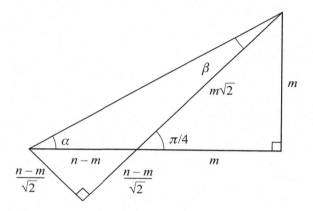

$$\alpha = \tan^{-1}\left(\frac{m}{n}\right), \quad \beta = \tan^{-1}\left(\frac{(n-m)/\sqrt{2}}{(n-m)/\sqrt{2}+m\sqrt{2}}\right) = \tan^{-1}\left(\frac{n-m}{n+m}\right)$$

$$\alpha + \beta = \pi/4$$

—Geoffrey A. Kandall

II.

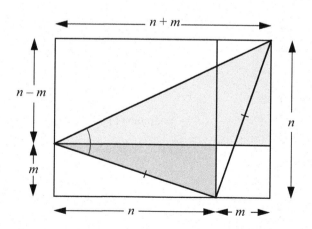

—RBN

One Figure, Five Arctangent Identities

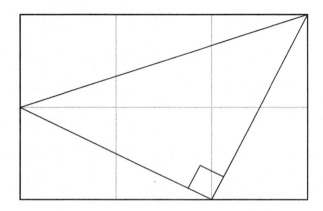

$$\frac{\pi}{4} = \arctan\left(\frac{1}{2}\right) + \arctan\left(\frac{1}{3}\right)$$

$$\frac{\pi}{4} = \arctan(3) - \arctan\left(\frac{1}{2}\right)$$

$$\frac{\pi}{4} = \arctan(2) - \arctan\left(\frac{1}{3}\right)$$

$$\frac{\pi}{2} = \arctan(1) + \arctan\left(\frac{1}{2}\right) + \arctan\left(\frac{1}{3}\right)$$

$$\pi = \arctan(1) + \arctan(2) + \arctan(3)$$

—Rex H. Wu

The Formulas of Hutton and Strassnitzky

Hutton's formula: $\quad \dfrac{\pi}{4} = 2\arctan\dfrac{1}{3} + \arctan\dfrac{1}{7}$ (1)

Strassnitzky's formula: $\quad \dfrac{\pi}{4} = \arctan\dfrac{1}{2} + \arctan\dfrac{1}{5} + \arctan\dfrac{1}{8}$ (2)

Proof.

(1)

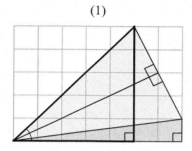

(2)

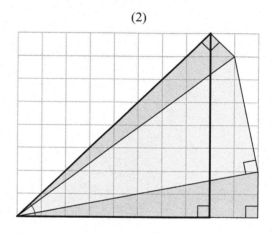

NOTE: Charles Hutton published (1) in 1776, and in 1789 Georg von Vega used it with Gregory's arctangent series to compute π to 143 decimal places, of which the first 126 were correct. L. K. Schulz von Strassnitzky provided (2) to Zacharias Dahse in 1844, who then used it to compute π correct to 200 decimal places.

—RBN

An Arctangent Identity

$$\arctan\left(x + \sqrt{1+x^2}\right) = \frac{\pi}{4} + \frac{1}{2}\arctan x$$

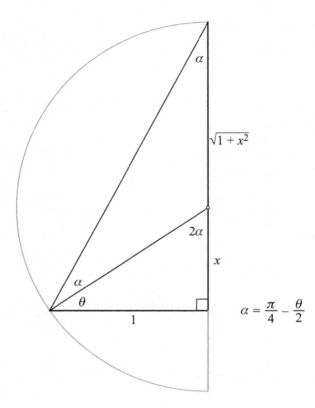

$$\alpha = \frac{\pi}{4} - \frac{\theta}{2}$$

—P. D. Barry

Euler's Arctangent Identity

$$\arctan\left(\frac{1}{x}\right) = \arctan\left(\frac{1}{x+y}\right) + \arctan\left(\frac{y}{x^2+xy+1}\right)$$

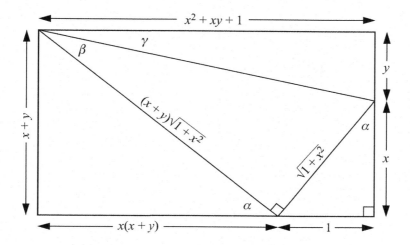

$$\alpha = \beta + \gamma.$$

NOTE: This is one of the many elegant arctangent identities discovered by Leonhard Euler. He employed them in the computation of π. For $x = y = 1$, we have Euler's Machin-like formula, $\pi/4 = \arctan(1/2) + \arctan(1/3)$. For $x = 2$ and $y = 1$, $\arctan(1/2) = \arctan(1/3) + \arctan(1/7)$. Substitute this into the previous identity, we obtain Hutton's formula, $\pi/4 = 2\arctan(1/3) + \arctan(1/7)$. In conjunction with the power series for arctangent, Hutton's formula was used as a check by Clausen in 1847 in computing π to 248 decimal places.

—Rex H. Wu

Extrema of the Function $a \cos t + b \sin t$

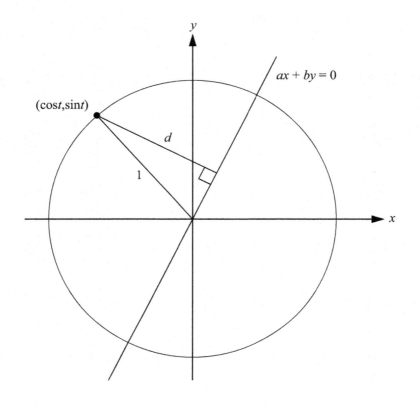

$$d \leq 1 \Rightarrow \frac{|a \cos t + b \sin t|}{\sqrt{a^2 + b^2}} \leq 1$$

$$-\sqrt{a^2 + b^2} \leq a \cos t + b \sin t \leq \sqrt{a^2 + b^2}$$

—M. Bayat, M. Hassani, & H. Teimoori

A Minimum Area Problem

For positive a and b, find the line through the point (a, b) that cuts off the triangle of smallest area K in the first quadrant.

Solution.

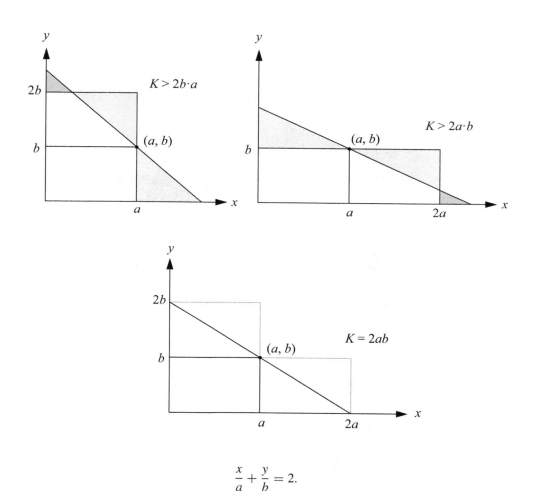

$$\frac{x}{a} + \frac{y}{b} = 2.$$

The Derivative of the Sine

$$\frac{d}{d\theta} \sin \theta = \cos \theta$$

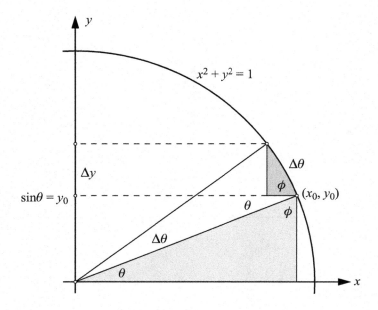

$$\frac{dy}{d\theta} \cong \frac{\Delta y}{\Delta \theta} = \frac{x_0}{1} = \sin \phi = \cos \theta.$$

—Donald Hartig

The Derivative of the Tangent

$$\frac{d}{d\theta}\tan\theta = \sec^2\theta$$

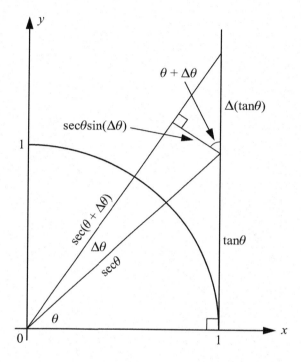

$$\frac{\sec(\theta + \Delta\theta)}{1} = \frac{\Delta(\tan\theta)}{\sec\theta\sin(\Delta\theta)}$$

$$\frac{\Delta(\tan\theta)}{\Delta\theta} = \sec\theta\sec(\theta + \Delta\theta)\frac{\sin(\Delta\theta)}{\Delta\theta}$$

$$\therefore \frac{d(\tan\theta)}{d\theta} = \sec^2\theta$$

—Yukio Kobayashi

Geometric Evaluation of a Limit II

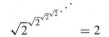

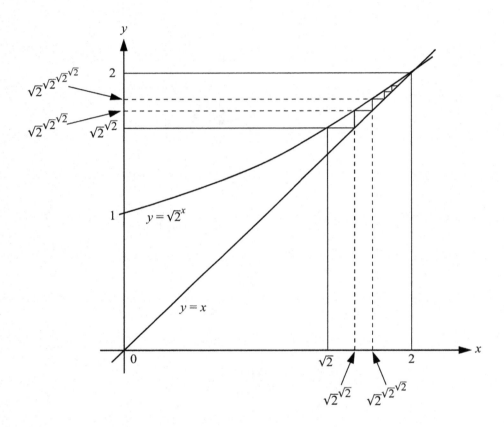

—F. Azarpanah

The Logarithm of a Number and Its Reciprocal

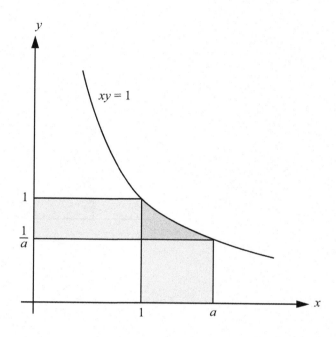

$$\int_{1/a}^{1} \frac{1}{y}\,dy = \int_{1}^{a} \frac{1}{x}\,dx$$

$$-\ln \frac{1}{a} = \ln a$$

—Vincent Ferlini

Regions Bounded by the Unit Hyperbola with Equal Area

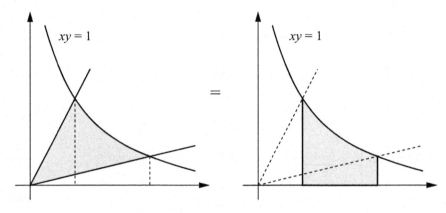

Proof.

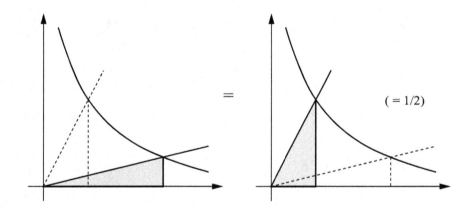

The Weierstrass Substitution II

(Karl Theodor Wilhelm Weierstrass, 1815–1897)

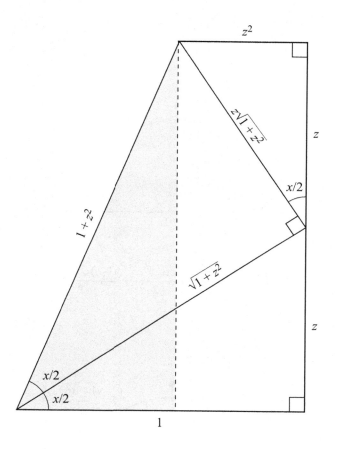

$$z = \tan \frac{x}{2} \quad \Rightarrow \quad \sin x = \frac{2z}{1 + z^2}, \quad \cos x = \frac{1 - z^2}{1 + z^2}.$$

—Sidney H. Kung

Look Ma, No Substitution!

$$\int_a^1 \sqrt{1 - x^2}\,dx = \frac{\cos^{-1} a}{2} - \frac{a\sqrt{1 - a^2}}{2}, \quad a \in [-1, 1].$$

I. $a \in [-1, 0]$

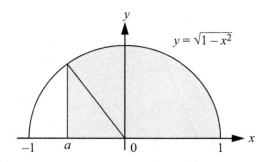

$$\int_a^1 \sqrt{1 - x^2}\,dx = \frac{\cos^{-1} a}{2} + \frac{(-a)\sqrt{1 - a^2}}{2}.$$

II. $a \in [0, 1]$

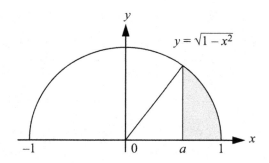

$$\int_a^1 \sqrt{1 - x^2}\,dx = \frac{\cos^{-1} a}{2} - \frac{a\sqrt{1 - a^2}}{2}.$$

—Marc Chamberland

Integrating the Natural Logarithm

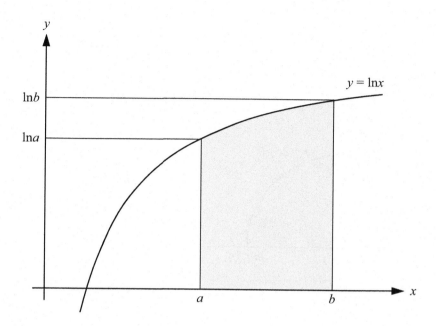

$$\int_a^b \ln x \, dx = b \ln b - a \ln a - \int_{\ln a}^{\ln b} e^y \, dy$$
$$= x \ln x |_a^b - (b - a)$$
$$= (x \ln x - x)|_a^b$$

—RBN

The Integrals of $\cos^2 \theta$ and $\sec^2 \theta$

I. $\int \cos^2 \theta \, d\theta = \frac{1}{2}\theta + \frac{1}{4}\sin 2\theta$

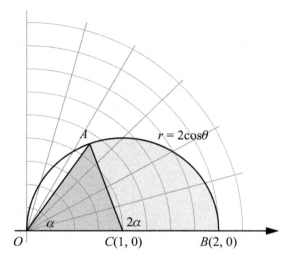

$$\int_0^\alpha \cos^2 \theta \, d\theta = \frac{1}{2} \int_0^\alpha \frac{1}{2} r^2 \, d\theta$$

$$= \frac{1}{2} \left(\text{Area} \triangle OAC + \text{Area Sector} ACB \right)$$

$$= \frac{1}{2} \left(\frac{1}{2} \cdot 1 \cdot \sin 2\alpha + \frac{1}{2} \cdot 1^2 \cdot 2\alpha \right)$$

$$= \frac{1}{2}\alpha + \frac{1}{4}\sin 2\alpha$$

II. $\int \sec^2 \theta \, d\theta = \tan \theta$

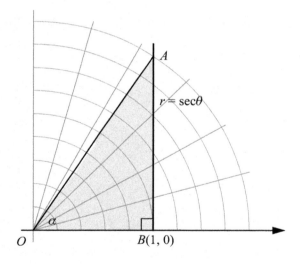

$$\int_0^\alpha \sec^2 \theta \, d\theta = 2 \int_0^\alpha \frac{1}{2} r^2 \, d\theta$$

$$= 2 \, \text{Area} \triangle OBA$$

$$= 2 \left(\frac{1}{2} \cdot 1 \cdot \tan \alpha \right)$$

$$= \tan \alpha$$

—Nick Lord

A Partial Fraction Decomposition

$$\frac{1}{n(n+1)} = \frac{1}{n} - \frac{1}{n+1}$$

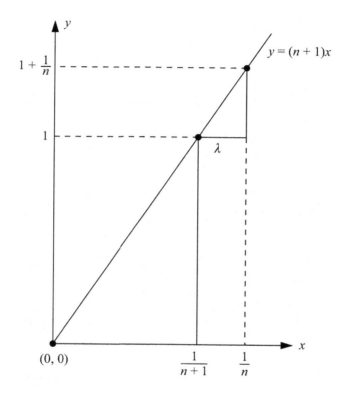

$$\lambda = \frac{1}{n} - \frac{1}{n+1}, \quad \frac{1/(n+1)}{1} = \frac{\lambda}{1/n} \quad \Rightarrow \quad \frac{1}{n} - \frac{1}{n+1} = \frac{1}{n} \cdot \frac{1}{n+1}.$$

—Steven J. Kifowit

An Integral Transform

$$\int_a^b f(x)dx = \int_a^b f(a+b-x)dx = \int_a^{(a+b)/2} (f(x)+f(a+b-x))\,dx$$

$$= \int_{(a+b)/2}^b (f(x)+f(a+b-x))\,dx$$

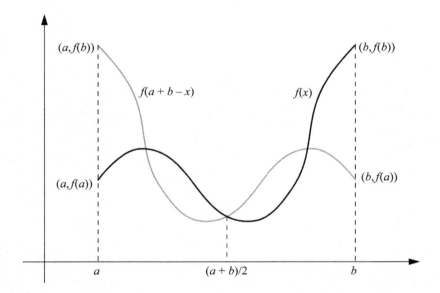

Example.

$$\int_0^{\pi/4} \ln(1+\tan x)dx = \int_0^{\pi/8} (\ln(1+\tan x)+\ln(1+\tan(\pi/4-x)))\,dx$$

$$= \int_0^{\pi/8} \left(\ln(1+\tan x)+\ln\left(1+\frac{1-\tan x}{1+\tan x}\right)\right)dx$$

$$= \int_0^{\pi/8} \ln 2\,dx = \frac{\pi}{8}\ln 2.$$

Exercises. (a) $\displaystyle\int_0^{\pi/2} \frac{dx}{1+\tan^\alpha x} = \frac{\pi}{4}$; (b) $\displaystyle\int_{-1}^1 \arctan(e^x)dx = \frac{\pi}{2}$;

(c) $\displaystyle\int_0^4 \frac{dx}{4+2^x} = \frac{1}{2}$; (d) $\displaystyle\int_0^{2\pi} \frac{dx}{1+e^{\sin x}} = \pi.$

—Sidney H. Kung

Inequalities

The Arithmetic Mean–Geometric Mean Inequality VII

$$a, b > 0 \quad \Rightarrow \quad \frac{a+b}{2} \geq \sqrt{ab}$$

I.

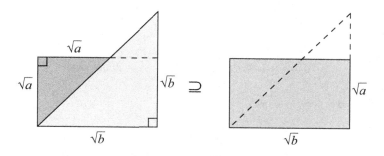

$$\frac{a}{2} + \frac{b}{2} \geq \sqrt{ab}.$$

—Edwin Beckenbach & Richard Bellman

II.

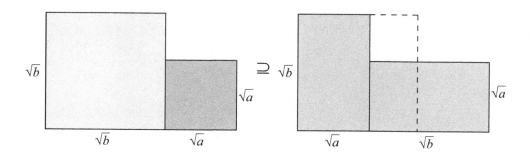

$$a + b \geq 2\sqrt{ab}.$$

—Alfinio Flores

The Arithmetic Mean–Geometric Mean Inequality VIII (via Trigonometry)

I. $x \in (0, \pi/2) \quad \Rightarrow \quad \tan x + \cot x \geq 2.$

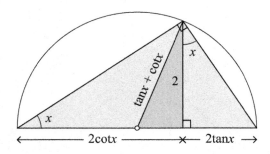

II. $a, b > 0 \quad \Rightarrow \quad \frac{a+b}{2} \geq \sqrt{ab}.$

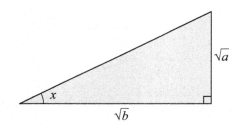

$$\frac{\sqrt{a}}{\sqrt{b}} + \frac{\sqrt{b}}{\sqrt{a}} \geq 2 \quad \Rightarrow \quad \frac{a+b}{2} \geq \sqrt{ab}.$$

—RBN

The Arithmetic Mean-Root Mean Square Inequality

$$a, b \geq 0 \quad \Rightarrow \quad \frac{a+b}{2} \leq \sqrt{\frac{a^2+b^2}{2}}$$

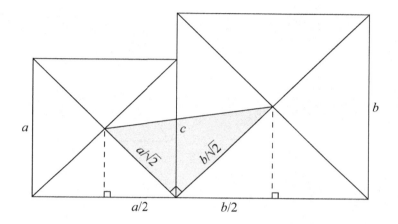

$$c^2 = \left(\frac{a}{\sqrt{2}}\right)^2 + \left(\frac{b}{\sqrt{2}}\right)^2 = \frac{a^2}{2} + \frac{b^2}{2},$$

$$\frac{a}{2} + \frac{b}{2} \leq c \quad \Rightarrow \quad \frac{a+b}{2} \leq \sqrt{\frac{a^2+b^2}{2}}.$$

—Juan-Bosco Romero Márquez

The Cauchy-Schwarz Inequality II
(via Pappus' theorem*)

(Augustin-Louis Cauchy, 1789–1857; Hermann Amandus Schwarz, 1843–1921)

$$|ax + by| \le \sqrt{a^2 + b^2}\sqrt{x^2 + y^2}$$

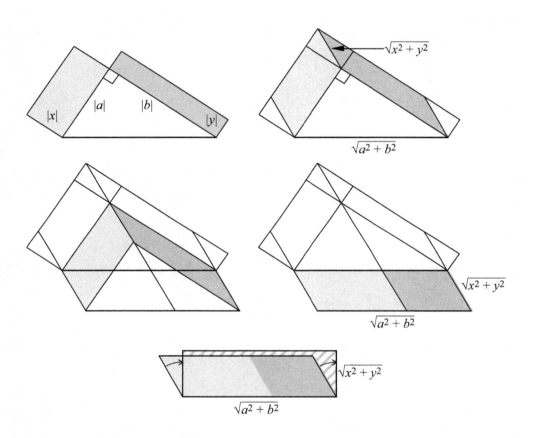

$$|ax + by| \le |a|\,|x| + |b|\,|y| \le \sqrt{a^2 + b^2}\sqrt{x^2 + y^2}.$$

*See p. 7.

—Claudi Alsina

The Cauchy-Schwarz Inequality III

$$|ax + by| \leq \sqrt{a^2 + b^2}\sqrt{x^2 + y^2}$$

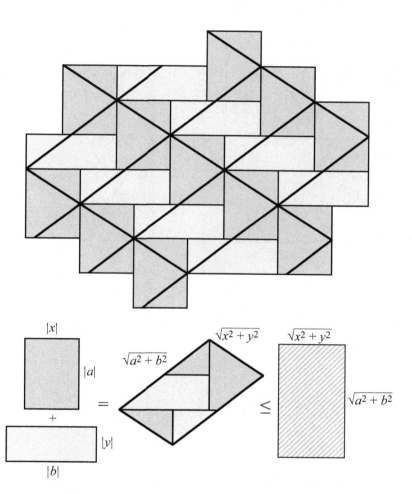

$$|ax + by| \leq |a|\,|x| + |b|\,|y| \leq \sqrt{a^2 + b^2}\sqrt{x^2 + y^2}.$$

—RBN

The Cauchy-Schwarz Inequality IV

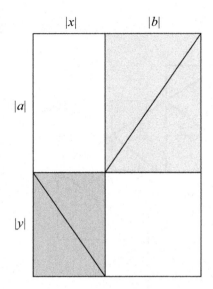 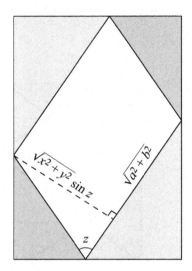

$$|a|\,|x| + |b|\,|y| = \sqrt{a^2 + b^2}\sqrt{x^2 + y^2}\,\sin z$$
$$\Rightarrow |\langle a, b\rangle \cdot \langle x, y\rangle| \le \|\langle a, b\rangle\|\,\|\langle x, y\rangle\|.$$

—Sidney H. Kung

The Cauchy-Schwarz Inequality V

$$|ax + by| \leq \sqrt{a^2 + b^2}\sqrt{x^2 + y^2}$$

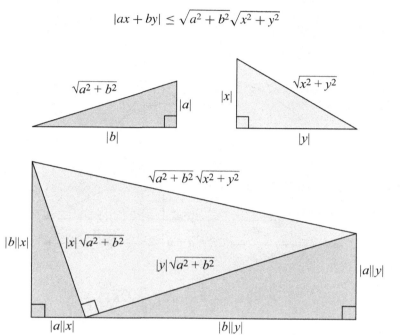

$$|ax + by| \leq |a|\,|x| + |b|\,|y| \leq \sqrt{a^2 + b^2}\sqrt{x^2 + y^2}.$$

—Claudi Alsina & RBN

Inequalities for the Radii of Right Triangles

If r, R, and K denote the inradius, circumradius, and area, respectively, of a right triangle, then

I. $R + r \geq \sqrt{2K}$.

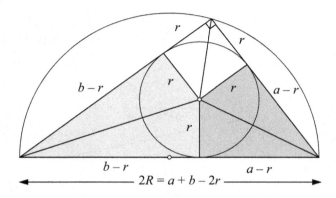

$$R + r = \frac{a+b}{2} \geq \sqrt{ab} = \sqrt{2K}.$$

II. $R/r \geq \sqrt{2} + 1$.

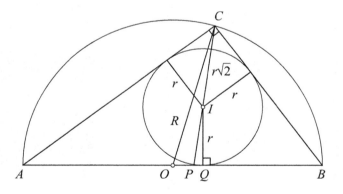

$$R = \overline{OC} \geq \overline{PC} \geq \overline{IC} + \overline{IQ} = r\sqrt{2} + r.$$

NOTE: For general triangles, the inequalities are $R + r \geq \sqrt{K\sqrt{3}}$ and $R/r \geq 2$, respectively.

Ptolemy's Inequality

In a convex quadrilateral with sides of length a, b, c, d (in that order) and diagonals of length p and q, we have $pq \leq ac + bd$.

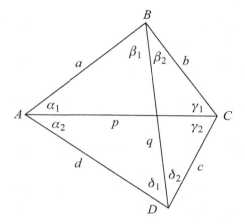

Proof.

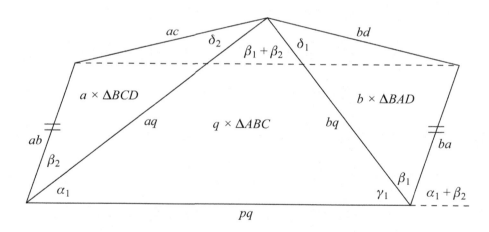

NOTE: The angle at the top of the figure $\delta_2 + \beta_1 + \beta_2 + \delta_1$ is drawn as being smaller than π, but the broken line representing $ac + bd$ is at least as long as the base of the parallelogram in any case. In a cyclic quadrilateral we have *Ptolemy's theorem*, see pp. 22–23.

—Claudi Alsina & RBN

An Algebraic Inequality I

(Problem 4, 2010 Kazakh National Mathematical Olympiad Final Round)

For $x, y \geq 0$ prove the inequality

$$\sqrt{x^2 - x + 1}\sqrt{y^2 - y + 1} + \sqrt{x^2 + x + 1}\sqrt{y^2 + y + 1} \geq 2(x + y).$$

Solution. (via Ptolemy's inequality):

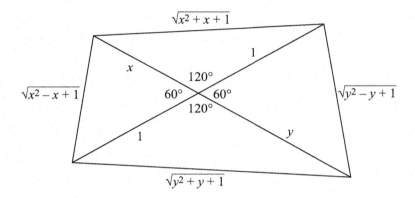

$$\sqrt{x^2 - x + 1}\sqrt{y^2 - y + 1} + \sqrt{x^2 + x + 1}\sqrt{y^2 + y + 1} \geq 2(x + y).$$

—Madeubek Kungozhin & Sidney H. Kung

An Algebraic Inequality II

(Problem 12, 1989 Leningrad Mathematics Olympiad, Grade 7, Second Round)

Let $a \geq b \geq c \geq 0$, and let $a + b + c \leq 1$. Prove $a^2 + 3b^2 + 5c^2 \leq 1$.

Solution.

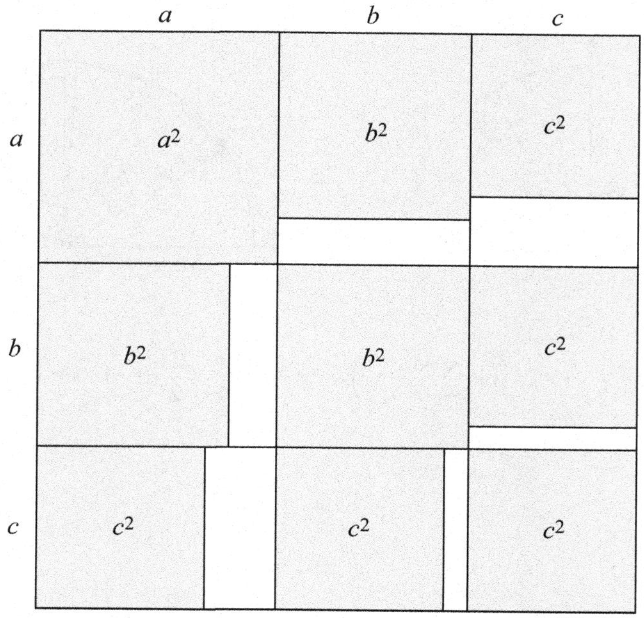

$$a^2 + 3b^2 + 5c^2 \leq (a + b + c)^2 \leq 1.$$

—Wei-Dong Jiang

The Sine is Subadditive on $[0, \pi]$

If $x_k \geq 0$ for $k = 1, 2, \ldots, n$ and $\sum_{k=1}^{n} x_k \leq \pi$, then

$$\sin\left(\sum_{k=1}^{n} x_k\right) \leq \sum_{k=1}^{n} \sin x_k.$$

Proof.

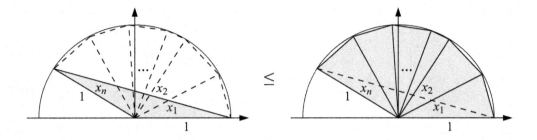

$$\frac{1}{2} \cdot 1 \cdot 1 \cdot \sin\left(\sum_{k=1}^{n} x_k\right) \quad \leq \quad \sum_{k=1}^{n} \frac{1}{2} \cdot 1 \cdot 1 \cdot \sin x_k$$

—Xingya Fan

The Tangent is Superadditive on $[0, \pi/2)$

If $x_k \geq 0$ for $k = 1, 2, \ldots, n$ and $\sum_{k=1}^{n} x_k < \pi/2$, then

$$\tan\left(\sum_{k=1}^{n} x_k\right) \geq \sum_{k=1}^{n} \tan x_k.$$

Proof.

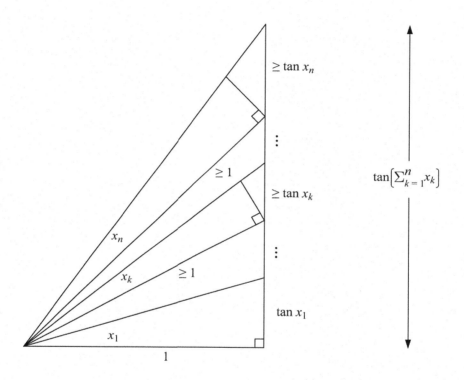

—Rob Pratt

Inequalities for Two Numbers whose Sum is One

$$p, q > 0, \ p + q = 1 \quad \Rightarrow \quad \frac{1}{p} + \frac{1}{q} \geq 4 \text{ and } \left(p + \frac{1}{p}\right)^2 + \left(q + \frac{1}{q}\right)^2 \geq \frac{25}{2}$$

(a)

(b)

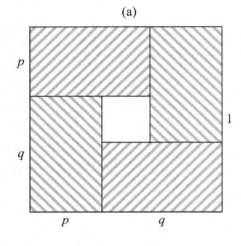
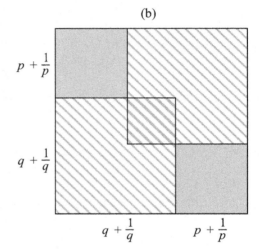

(a) $1 \geq 4pq \quad \Rightarrow \quad \frac{1}{p} + \frac{1}{q} \geq 4.$

(b) $2\left(p + \frac{1}{p}\right)^2 + 2\left(q + \frac{1}{q}\right)^2 \geq \left(p + \frac{1}{p} + q + \frac{1}{q}\right)^2 \geq (1+4)^2 = 25.$

—Claudi Alsina & RBN

Padoa's Inequality

(Alessandro Padoa, 1868–1937)

If a, b, c are the sides of a triangle, then

$$abc \geq (a+b-c)(b+c-a)(c+a-b).$$

1.

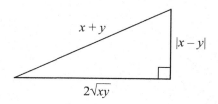

$$x + y \geq 2\sqrt{xy}.$$

2.

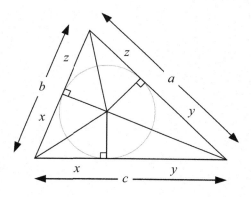

$$\begin{aligned}
abc &= (y+z)(z+x)(x+y) \\
&\geq 2\sqrt{yz} \cdot 2\sqrt{zx} \cdot 2\sqrt{xy} \\
&= (2z)(2x)(2y) \\
&= (a+b-c)(b+c-a)(c+a-b).
\end{aligned}$$

—RBN

Steiner's Problem on the Number e

(Jakob Steiner, 1796–1863)

For what positive x is the xth root of x the greatest?

Solution. $x > 0 \implies \sqrt[x]{x} \le \sqrt[e]{e}.$

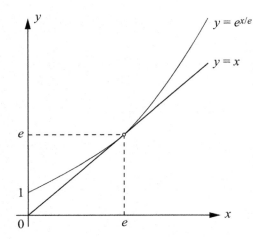

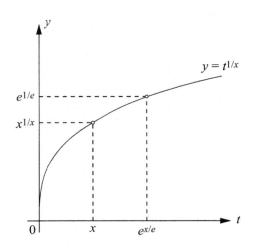

$$x \le e^{x/e} \qquad\qquad x^{1/x} \le e^{1/e}$$

[In the right-hand figure, $x > 1$; the other case differs only in concavity.]

Corollary. $e^\pi > \pi^e.$

—RBN

Simpson's Paradox

(Edward Hugh Simpson, 1922–)

1. Popularity of a candidate is greater among women than men in each town, yet popularity of the candidate in the whole district is greater among men.
2. Procedure X has greater success than procedure Y in each hospital, yet in general, procedure Y has greater success than X.

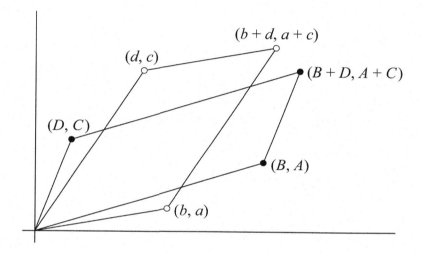

$$\frac{a}{b} < \frac{A}{B} \text{ and } \frac{c}{d} < \frac{C}{D}, \text{ yet } \frac{a+c}{b+d} > \frac{A+C}{C+D}.$$

1. In town 1, $B =$ the number of women, $b =$ the number of men, $A =$ the number of women favoring the candidate, $a =$ the number of men favoring the candidate; and similarly for town 2 with D, d, C, and c.
2. In hospital 1, $B =$ the number of patients treated with X, $b =$ the number of patients treated with Y, $A =$ the number of successful procedures with X, $a =$ the number of successful procedures with Y; and similarly for hospital 2 with D, d, C, and c.

—Jerzy Kocik

Markov's Inequality

(Andrei Andreyevich Markov, 1856–1922)

$$P[X \geq a] \leq \frac{E(X)}{a}$$

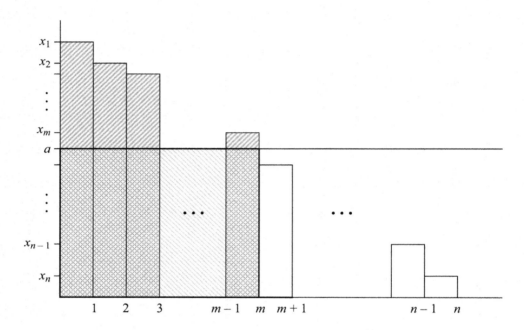

$$x_m \geq a \Rightarrow ma \leq \sum_{i=1}^{n} x_i \Rightarrow \frac{m}{n} \leq \frac{1}{a}\left(\frac{\sum_{i=1}^{n} x_i}{n}\right),$$

$$\therefore P[X \geq a] \leq \frac{E[X]}{a}.$$

—Pat Touhey

Integers
&
Integer Sums

Sums of Odd Integers IV

$$1 + 3 + 5 + \cdots + (2n - 1) = n^2$$

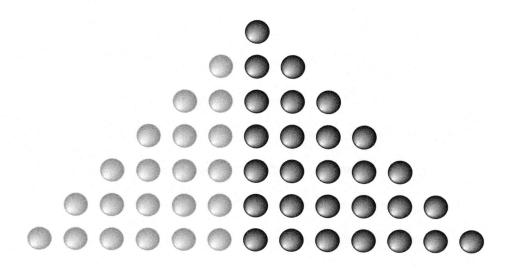

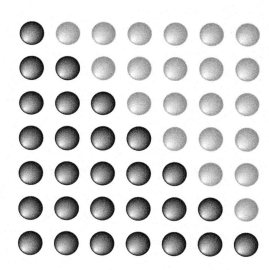

Sums of Odd Integers V

$$1 + 3 + 5 + \cdots + (2n-1) = n^2$$

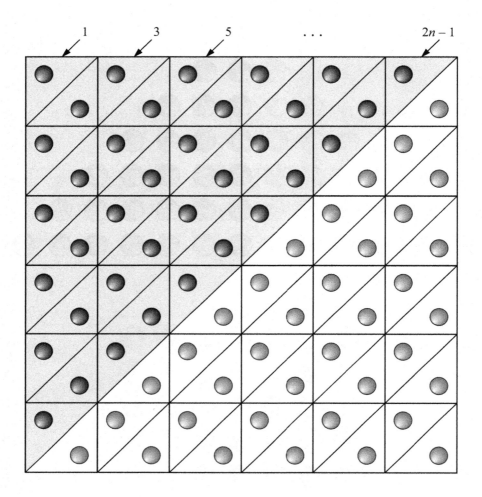

$$2\,[1 + 3 + 5 + \cdots + (2n-1)] = 2n^2.$$

—Timothée Duval

Alternating Sums of Odd Numbers

$$\sum_{k=1}^{n} (2k-1)(-1)^{n-k} = n$$

n odd:

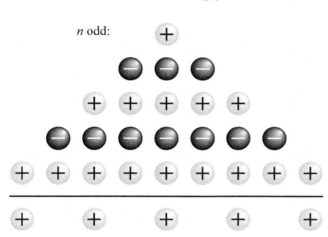

n even:

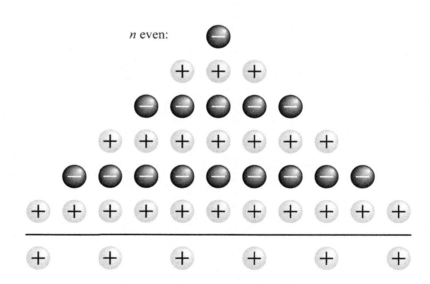

—Arthur T. Benjamin

Sums of Squares X

$$1^2 + 2^2 + \cdots + n^2 = \frac{1}{6}n(n+1)(2n+1)$$

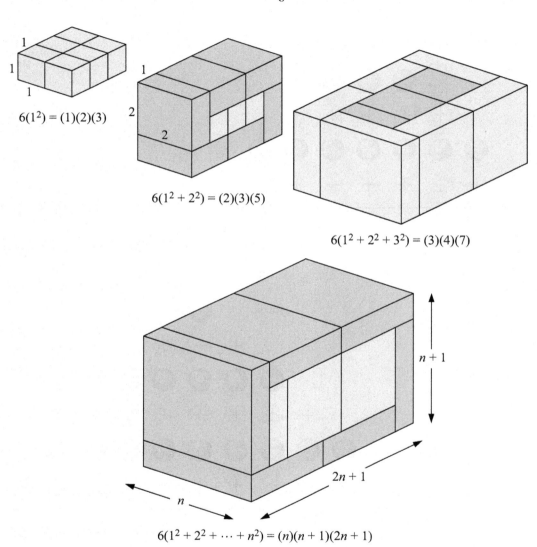

$6(1^2) = (1)(2)(3)$

$6(1^2 + 2^2) = (2)(3)(5)$

$6(1^2 + 2^2 + 3^2) = (3)(4)(7)$

$6(1^2 + 2^2 + \cdots + n^2) = (n)(n+1)(2n+1)$

NOTE: For a four-dimensional illustration of the sum of cubes formula, see Sasho Kala-jdzievski, Some evident summation formulas, *Math. Intelligencer* **22** (2000), pp. 47–49.

—Sasho Kalajdzievski

Sums of Squares XI

$$\sum_{k=1}^{n} k^2 = \sum_{i=1}^{n} \sum_{j=1}^{n} \min(i, j)$$

$$\sum_{k=1}^{n} k^2$$

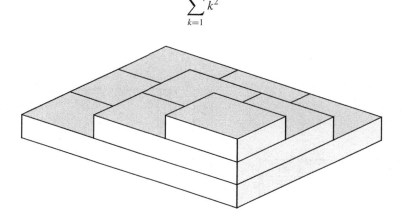

$$\sum_{i=1}^{n} \sum_{j=1}^{n} \min(i, j)$$

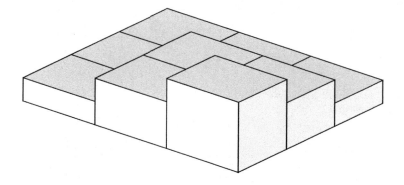

—Abraham Arcavi & Alfinio Flores

Alternating Sums of Consecutive Squares

$$2^2 - 3^2 + 4^2 = -5^2 + 6^2$$

$$4^2 - 5^2 + 6^2 - 7^2 + 8^2 = -9^2 + 10^2 - 11^2 + 12^2$$

$$6^2 - 7^2 + 8^2 - 9^2 + 10^2 - 11^2 + 12^2 = -13^2 + 14^2 - 15^2 + 16^2 - 17^2 + 18^2$$

$$\vdots$$

$$(2n)^2 - (2n+1)^2 + \cdots + (4n)^2 = -(4n+1)^2 + (4n+2)^2 - \cdots + (6n)^2$$

E.g., for $n = 2$:

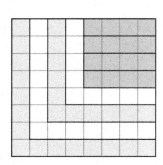 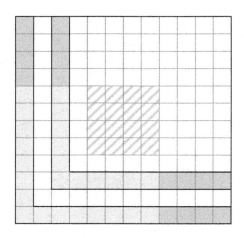

$$8^2 - 7^2 + 6^2 - 5^2 + 4^2 \qquad = \qquad 12^2 - 11^2 + 10^2 - 9^2.$$

Exercise. Show that

$$3^2 = -4^2 + 5^2$$

$$5^2 - 6^2 + 7^2 = -8^2 + 9^2 - 10^2 + 11^2$$

$$7^2 - 8^2 + 9^2 - 10^2 + 11^2 = -12^2 + 13^2 - 14^2 + 15^2 - 16^2 + 17^2$$

$$\vdots$$

$$(2n+1)^2 - (2n+2)^2 + \cdots + (4n-1)^2 = -(4n)^2 + (4n+1)^2 - \cdots + (6n-1)^2$$

—RBN

Alternating Sums of Squares of Odd Numbers

If n is even, $\displaystyle\sum_{k=1}^{n} (2k-1)^2(-1)^k = 2n^2$, e.g., $n = 4$:

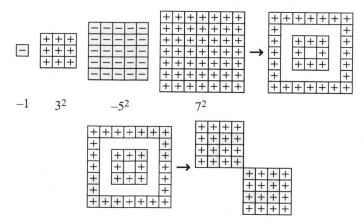

$$-1 \qquad 3^2 \qquad -5^2 \qquad 7^2$$

If n is odd, $\displaystyle\sum_{k=1}^{n} (2k-1)^2(-1)^{k-1} = 2n^2 - 1$, e.g., $n = 5$:

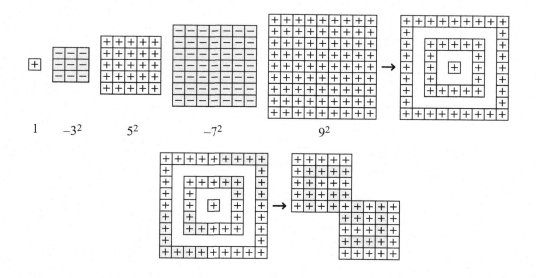

$$1 \qquad -3^2 \qquad 5^2 \qquad -7^2 \qquad 9^2$$

—Ángel Plaza

Archimedes' Sum of Squares Formula

$$3 \sum_{i=1}^{n} i^2 = (n+1)n^2 + \sum_{i=1}^{n} i$$

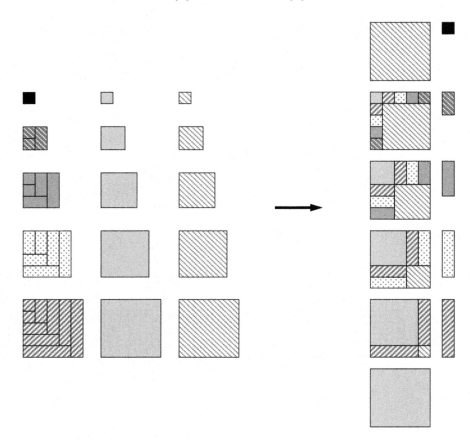

—Katherine Kanim

Summing Squares by Counting Triangles

$$(a+b+c)^2 + (a+b-c)^2 + (a-b+c)^2 + (-a+b+c)^2$$
$$= 4(a^2 + b^2 + c^2)$$

Proof by inclusion-exclusion, where each $\triangle$ or $\nabla = 1$, e.g., for $(a, b, c) = (5, 6, 7)$:

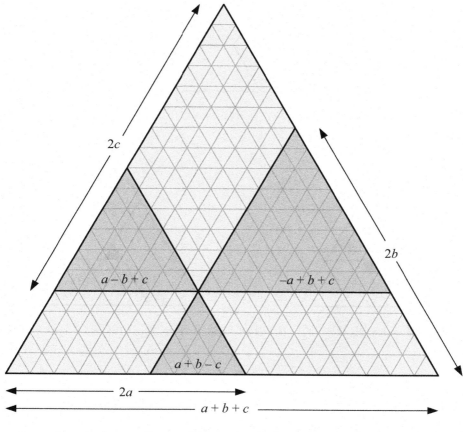

$$(a+b+c)^2 = (2a)^2 + (2b)^2 + (2c)^2 - (a+b-c)^2 - (a-b+c)^2 - (-a+b+c)^2.$$

—RBN

Squares Modulo 3

$$n^2 = 1 + 3 + 5 + \cdots + (2n-1) \Rightarrow \quad n^2 \equiv \begin{cases} 0(\text{mod}3), & n \equiv 0(\text{mod}3) \\ 1(\text{mod}3), & n \equiv \pm1(\text{mod}3) \end{cases}$$

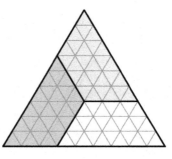

$$(3k)^2 = 3[(2k)^2 - k^2]$$

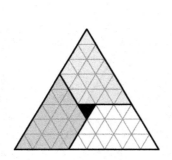

$$(3k-1)^2 = 1 + 3[(2k-1)^2 - (k-1)^2] \qquad (3k+1)^2 = 1 + 3[(2k+1)^2 - (k+1)^2]$$

—RBN

The Sum of Factorials of Order Two

$$1 \cdot 2 + 2 \cdot 3 + 3 \cdot 4 + \cdots + n(n+1) = \frac{n(n+1)(n+2)}{3}$$

1.

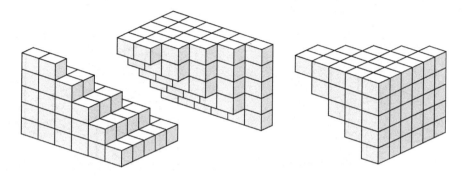

$$3 \cdot [1.2 + 2.3 + 3.4 + \cdots + n(n+1)].$$

2.

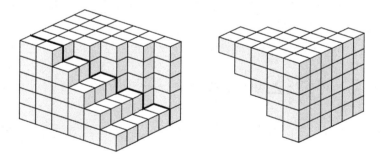

3.

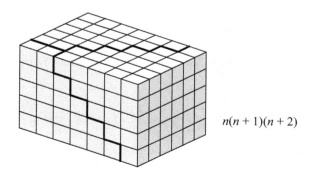

$n(n+1)(n+2)$

—Giorgio Goldoni

The Cube as a Double Sum

$$\sum_{i=1}^{n}\sum_{j=1}^{n}(i+j-1) = n^3$$

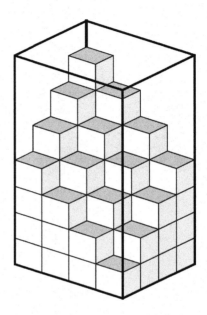

$$S = \sum_{i=1}^{n}\sum_{j=1}^{n}(i+j-1) \quad \Rightarrow \quad 2S = n^2 \cdot 2n = 2n^3.$$

NOTE: A similar figure yields the following result for sums of two-dimensional arithmetic progressions:

$$\sum_{i=1}^{m}\sum_{j=1}^{n}[a+(i-1)b+(j-1)c] = \frac{mn}{2}[2a+(m-1)b+(n-1)c].$$

As with one-dimensional arithmetic progressions, the sum is the number of terms times the average of the first $[(i, j) = (1, 1)]$ and last $[(i, j) = (m, n)]$ terms.

—RBN

The Cube as an Arithmetic Sum II

$$1 = 1$$
$$8 = 3 + 5$$
$$27 = 6 + 9 + 12$$
$$64 = 10 + 14 + 18 + 22$$
$$\vdots$$

$$t_n = 1 + 2 + \cdots + n \Rightarrow n^3 = t_n + (t_n + n) + (t_n + 2n) + \cdots + (t_n + (n-1)n)$$

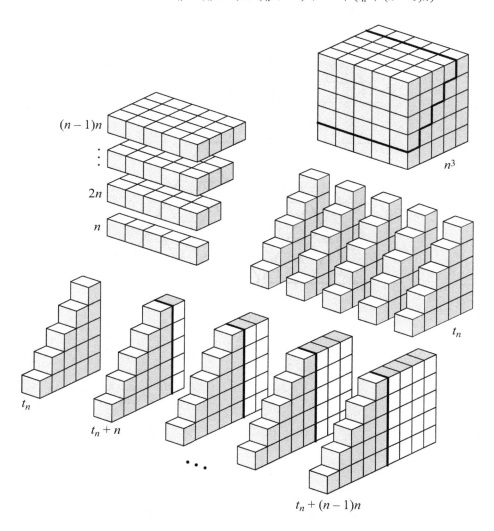

—RBN

Sums of Cubes VIII

$$1^3 + 2^3 + 3^3 + \cdots + n^3 = (1 + 2 + 3 + \cdots + n)^2$$

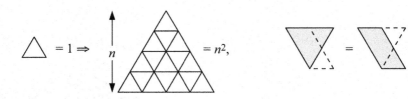

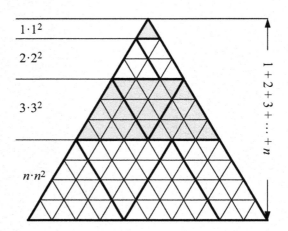

$$1^3 + 2^3 + 3^3 + \cdots + n^3 = (1 + 2 + 3 + \cdots + n)^2.$$

—Parames Laosinchai

The Difference of Consecutive Integer Cubes is Congruent to 1 Modulo 6

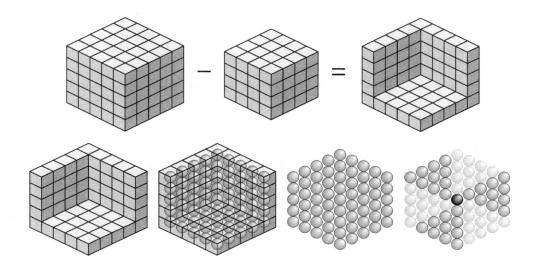

$$(n + 1)^3 - n^3 \equiv 1 (\mathrm{mod}\, 6).$$

—Claudi Alsina, Hasan Unal, & RBN

Fibonacci Identities II

(Leonardo of Pisa, circa 1170–1250)

$$F_1 = F_2 = 1, \quad F_n = F_{n-1} + F_{n-2} \Rightarrow$$

I. (a) $F_1F_2 + F_2F_3 + \cdots + F_{2n}F_{2n+1} = F_{2n+1}^2 - 1$,
   (b) $F_1F_2 + F_2F_3 + \cdots + F_{2n-1}F_{2n} = F_{2n}^2$.

(a) 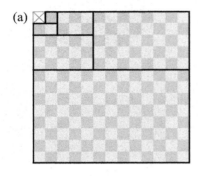   (b)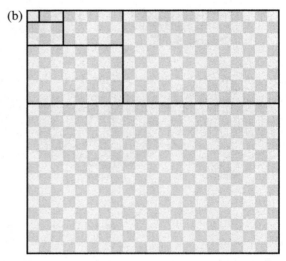

II. $F_1F_3 + F_2F_4 + \cdots + F_{2n}F_{2n+2} = F_2^2 + F_3^2 + \cdots F_{2n+1}^2$
$$= F_{2n+1}F_{2n+2} - 1.$$

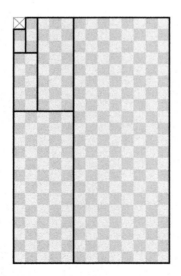 =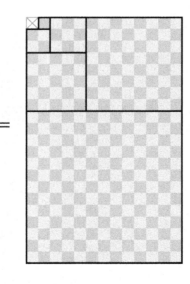

Fibonacci Tiles

$$F_0 = 0, \quad F_1 = 1, \quad F_{n+1} = F_n + F_{n-1} \implies$$

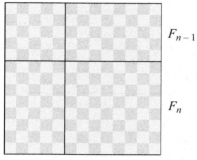

$$\begin{aligned}
F_{n+1}^2 &= 2F_{n+1}F_n - F_n^2 + F_{n-1}^2 \\
&= 2F_{n+1}F_{n-1} + F_n^2 - F_{n-1}^2 \\
&= 2F_nF_{n-1} + F_n^2 + F_{n-1}^2 \\
&= F_{n+1}F_n + F_nF_{n-1} + F_{n-1}^2 \\
&= F_{n+1}F_{n-1} + F_n^2 + F_nF_{n-1}
\end{aligned}$$

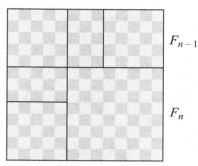

$$F_{n+1}^2 = F_n^2 + 3F_{n-1}^2 + 2F_{n-1}F_{n-2}$$

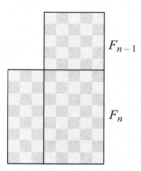

$$F_n^2 = F_{n+1}F_{n-1} + F_nF_{n-2} - F_{n-1}^2$$

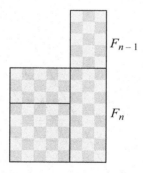

$$F_n^2 = F_{n+1}F_{n-2} + F_{n-1}^2$$

—Richard L. Ollerton

Fibonacci Trapezoids

I. Recursion: $F_n + F_{n+1} = F_{n+2}$.

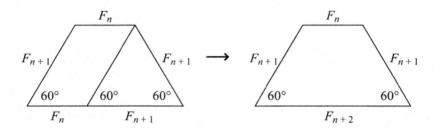

II. Identity: $1 + \sum_{k=1}^{n} F_k = F_{n+2}$.

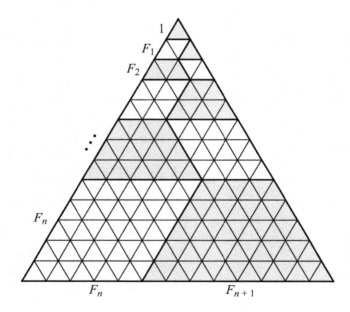

—Hans Walser

Fibonacci Triangles and Trapezoids

$$F_1 = F_2 = 1, \quad F_{n+2} = F_{n+1} + F_n \quad \Rightarrow \quad \sum_{k=1}^{n} F_k^2 = F_n F_{n+1}$$

I. Counting triangles:

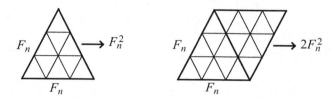

II. Identity: $F_n^2 + F_{n+1}^2 + \sum_{k=1}^{n} 2F_k^2 = (F_n + F_{n+1})^2$:

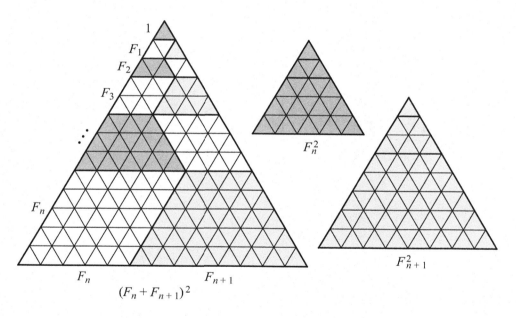

III. $\therefore \sum_{k=1}^{n} F_k^2 = F_n F_{n+1}$.

—Ángel Plaza & Hans Walser

Fibonacci Squares and Cubes

$$F_1 = F_2 = 1, \quad F_n = F_{n-1} + F_{n-2} \quad \Rightarrow$$

I. $F_{n+1}^2 = F_n^2 + F_{n-1}^2 + 2F_{n-1}F_n.$

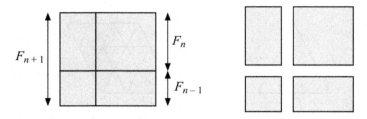

II. $F_{n+1}^3 = F_n^3 + F_{n-1}^3 + 3F_{n-1}F_nF_{n+1}.$

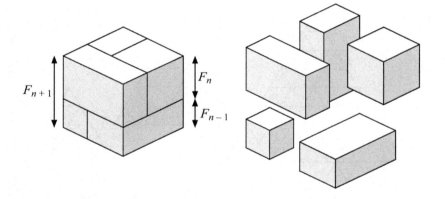

QUERY: Is there an analogous result in four dimensions?

—RBN

Every Octagonal Number is the Difference of Two Squares

$$1 = 1 = 1^2 - 0^2$$
$$1 + 7 = 8 = 3^2 - 1^2$$
$$1 + 7 + 13 = 21 = 5^2 - 2^2$$
$$1 + 7 + 13 + 19 = 40 = 7^2 - 3^2$$
$$\vdots$$
$$O_n = 1 + 7 + \cdots + (6n - 5) = (2n - 1)^2 - (n - 1)^2$$

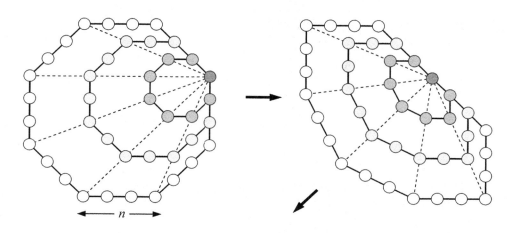

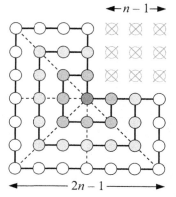

—RBN

Powers of Two

$$1 + 1 + 2 + 2^2 + \cdots + 2^{n-1} = 2^n.$$

—James Tanton

Sums of Powers of Four

$$\sum_{k=0}^{n} 4^k = \frac{4^{n+1} - 1}{3}$$

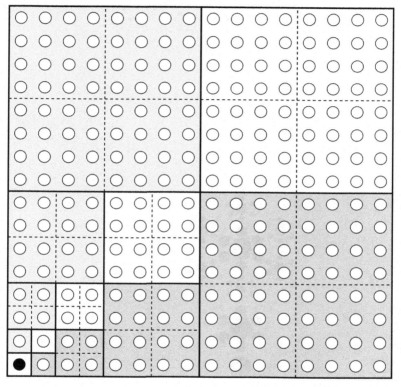

$$1 + 1 + 2 + 4 + \cdots + 2^n = 2^{n+1}$$

$$1 + 3\left(1 + 4 + 4^2 + \cdots + 4^n\right) = \left(2^{n+1}\right)^2 = 4^{n+1}.$$

—David B. Sher

Sums of Consecutive Powers of n via Self-Similarity

For any integers $n \geq 4$ and $k \geq 0$

$$1 + n + n^2 + \cdots + n^k = \frac{n^{k+1} - 1}{n - 1}.$$

E.g., $n = 7, k = 2$:

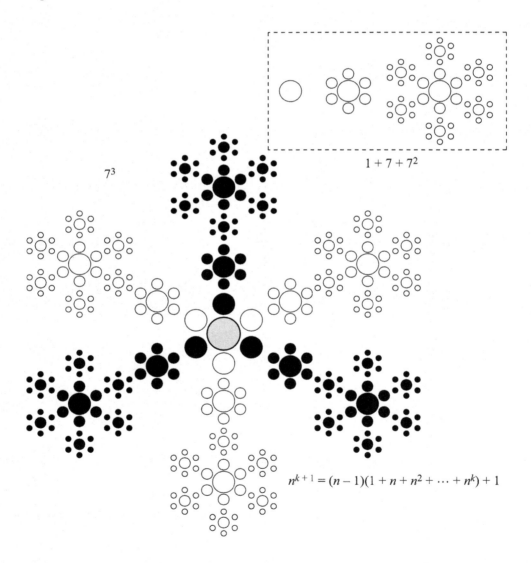

$1 + 7 + 7^2$

7^3

$$n^{k+1} = (n-1)(1 + n + n^2 + \cdots + n^k) + 1$$

—Mingjang Chen

Every Fourth Power Greater than One is the Sum of Two Non-consecutive Triangular Numbers

$$t_k = 1 + 2 + \cdots + k \quad \Rightarrow \quad 2^4 = 15 + 1 = t_5 + t_1,$$

$$3^4 = 66 + 15 = t_{11} + t_5,$$

$$4^4 = 190 + 66 = t_{19} + t_{11},$$

$$\vdots$$

$$n^4 = t_{n^2+n-1} + t_{n^2-n-1}.$$

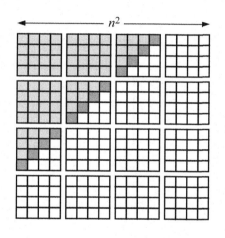

$$n(n-1) \,\blacksquare\, = 2\frac{n(n-1)}{2} = 2t_{n-1}$$

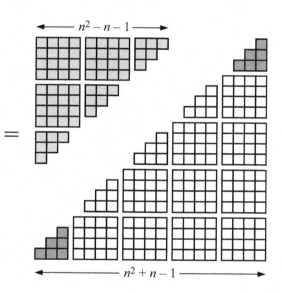

NOTE: Since $k^2 = t_{k-1} + t_k$, we also have $n^4 = t_{n^2-1} + t_{n^2}$.

—RBN

Sums of Triangular Numbers V

$$t_k = 1 + 2 + \cdots + k \quad \Rightarrow \quad t_1 + t_2 + \cdots + t_n = \frac{n(n+1)(n+2)}{6}$$

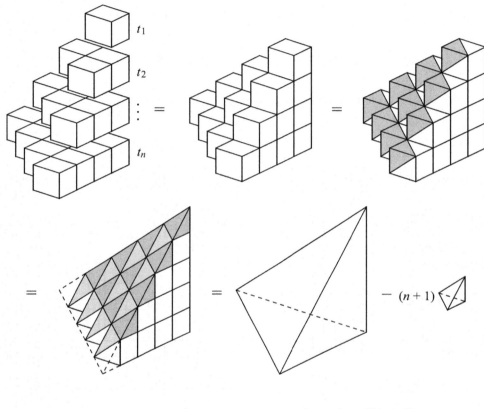

$$t_1 + t_2 + \cdots + t_n = \frac{1}{6}(n+1)^3 - (n+1) \cdot \frac{1}{6} = \frac{n(n+1)(n+2)}{6}.$$

—RBN

Alternating Sums of Triangular Numbers II

$$t_k = 1 + 2 + \cdots + k \quad \Rightarrow \quad \sum_{k=1}^{2n}(-1)^k t_k = 2t_n$$

E.g., $n = 3$:

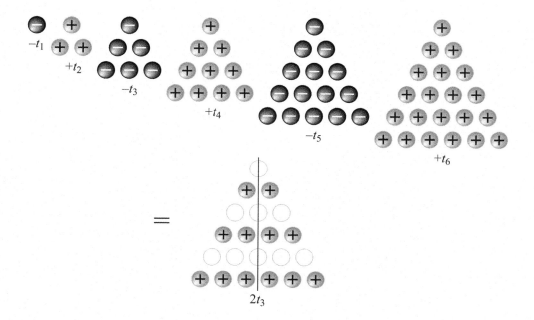

—Ángel Plaza

Runs of Triangular Numbers

$$t_k = 1 + 2 + \cdots + k \quad \Rightarrow$$

$$t_1 + t_2 + t_3 = t_4$$

$$t_5 + t_6 + t_7 + t_8 = t_9 + t_{10}$$

$$t_{11} + t_{12} + t_{13} + t_{14} + t_{15} = t_{16} + t_{17} + t_{18}$$

$$\vdots$$

$$t_{n^2-n-1} + t_{n^2-n} + \cdots + t_{n^2-1} = t_{n^2} + t_{n^2+1} + \cdots + t_{n^2+n-2}.$$

E.g., $n = 4$:

I. $t_{16} + t_{17} + t_{18} = t_{15} + t_{14} + t_{13} + 1 \cdot 4^2 + 3 \cdot 4^2 + 5 \cdot 4^2$;

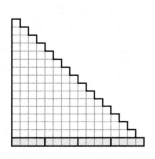 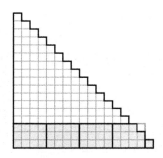 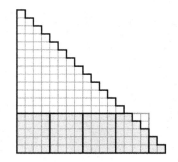

II. $(1 + 3 + 5) \cdot 4^2 = 12^2 = t_{12} + t_{11}$;

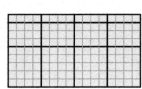 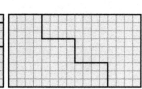 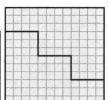 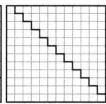

III. $\therefore t_{11} + t_{12} + t_{13} + t_{14} + t_{15} = t_{16} + t_{17} + t_{18}.$

—Hasan Unal & RBN

Sums of Every Third Triangular Number

$$t_k = 1 + 2 + 3 + \cdots + k \quad \Rightarrow \quad t_3 + t_6 + t_9 + \cdots + t_{3n} = 3(n+1)t_n$$

I. $t_{3k} = 3(k^2 + t_k)$;

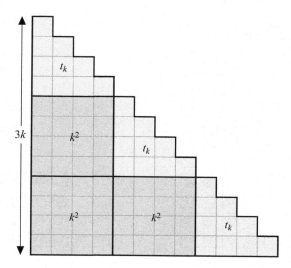

II. $\sum_{k=1}^{n} (k^2 + t_k) = (n+1)t_n$;

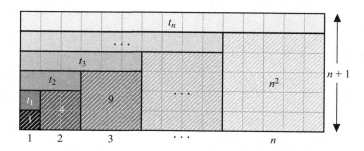

III. $\therefore \sum_{k=1}^{n} t_{3k} = 3(n+1)t_n$.

Exercise. Show that

$$t_2 + t_5 + t_8 + \cdots + t_{3n-1} = 3nt_n,$$
$$t_1 + t_4 + t_7 + \cdots + t_{3n-2} = 3(n-1)t_n + n.$$

—RBN

Triangular Sums of Odd Numbers

$$t_k = 1 + 2 + \cdots + k \quad \Rightarrow \quad \begin{cases} 1 + 5 + 9 + \cdots + (4n - 3) = t_{2n-1} \\ 3 + 7 + 11 + \cdots + (4n - 1) = t_{2n} \end{cases}$$

E.g., $n = 5$:

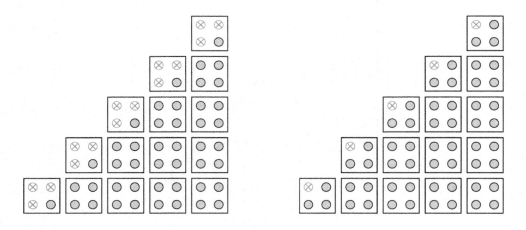

$$1 + 5 + 9 + 13 + 17 = 45 = t_9 \qquad\qquad 3 + 7 + 11 + 15 + 19 = 55 = t_{10}$$

—Yukio Kobayashi

Triangular Numbers are Binomial Coefficients

Lemma. There exists a one-to-one correspondence between a set with $t_n = 1 + 2 + \cdots + n$ elements and the set of two-element subsets of a set with $n + 1$ elements.

Proof.

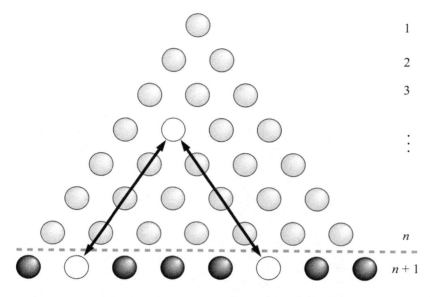

$$1$$
$$2$$
$$3$$
$$\vdots$$
$$n$$
$$n + 1$$

Theorem. $t_n = 1 + 2 + \cdots + n \quad \Rightarrow \quad t_n = \dbinom{n + 1}{2}.$

—Loren Larson

The Inclusion-Exclusion Formula for Triangular Numbers

Theorem. Let $t_k = 1 + 2 + \cdots + k$ and $t_0 = 0$. If $0 \leq a, b, c \leq n$ and $2n \leq a + b + c$, then

$$t_n = t_a + t_b + t_c - t_{a+b-n} - t_{b+c-n} - t_{c+a-n} + t_{a+b+c-2n}.$$

Proof.

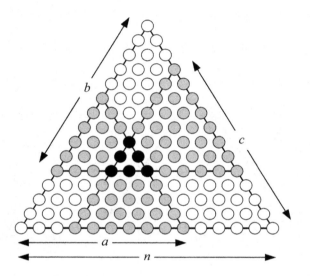

NOTES:

(1) If $0 \leq a, b, c \leq n$, $2n > a + b + c$, but $n \leq \min(a+b, b+c, c+a)$, then $t_n = t_a + t_b + t_c - t_{a+b-n} - t_{b+c-n} - t_{c+a-n} + t_{2n-a-b-c-1}$;

(2) the following special cases are of interest:

    (a) $(n; a, b, c) = (2n - k; k, k, k)$, $3(t_n - t_k) = t_{2n-k} - t_{2k-n}$;

    (b) $(n; a, b, c) = (a + b + c; 2a, 2b, 2c)$, $t_{2a} + t_{2b} + t_{2c} = t_{a+b+c} + t_{a+b-c} + t_{a-b+c} + t_{-a+b+c}$;

    (c) $(n; a, b, c) = (3k; 2k, 2k, 2k)$, $3(t_{2k} - t_k) = t_{3k}$.

—RBN

Partitioning Triangular Numbers

$$t_k = 1 + 2 + \cdots + k, \quad 1 \le q \le (n+1)/2 \quad \Rightarrow$$

1. $\quad t_n = 3t_q + 3t_{q-1} + 3t_{n-2q} - 2t_{n-3q}, \qquad n - 3q \ge 0;$

2. $\quad t_n = 3t_q + 3t_{q-1} + 3t_{n-2q} - 2t_{3q-n-1}, \quad n - 3q < 0.$

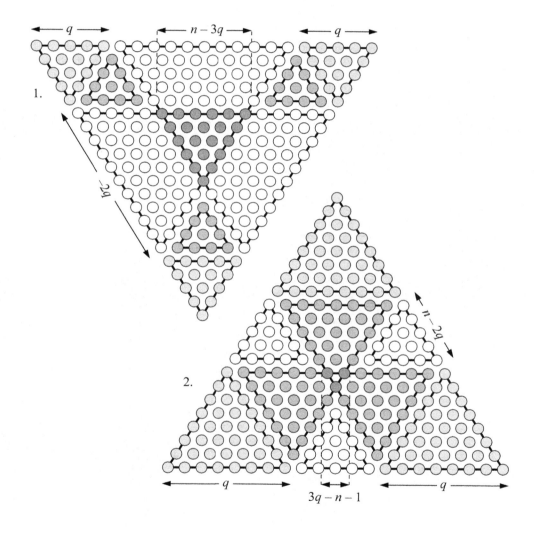

—Matthew J. Haines & Michael A. Jones

A Triangular Identity II

$$2 + 3 + 4 = 9 = 3^2 - 0^2$$
$$5 + 6 + 7 + 8 + 9 = 35 = 6^2 - 1^2$$
$$10 + 11 + 12 + 13 + 14 + 15 + 16 = 91 = 10^2 - 3^2$$
$$\vdots$$
$$t_n = 1 + 2 + \cdots + n \quad \Rightarrow \quad t_{(n+1)^2} - t_{n^2} = t_{n+1}^2 - t_{n-1}^2$$

E.g., $n = 4$:

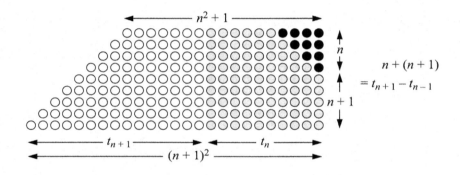

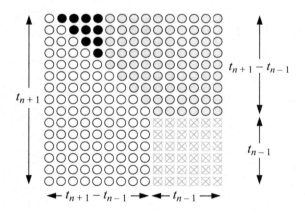

—RBN

A Triangular Sum

$$t(n) = 1 + 2 + \cdots + n \quad \Rightarrow \quad \sum_{k=0}^{n} t\left(2^k\right) = \frac{1}{3}t\left(2^{n+1} + 1\right) - 1$$

$$3\sum_{k=0}^{n} t\left(2^k\right) = t\left(2^{n+1} + 1\right) - 3.$$

Exercises. (a) $\sum_{k=1}^{n} t\left(2^k - 1\right) = \frac{1}{3}t\left(2^{n+1} - 2\right)$;

(b) $\sum_{k=0}^{n} t\left(3 \cdot 2^k - 1\right) = \frac{1}{3}\left[t\left(3 \cdot 2^{n+1} - 2\right) - 1\right]$.

—RBN

A Weighted Sum of Triangular Numbers

$$t_n = 1 + 2 + 3 + \cdots + n, \quad n \geq 1 \quad \Rightarrow$$

$$\sum_{k=1}^{n} k t_{k+1} = t_{t_{n+1}-1}.$$

E.g., $n = 4$:

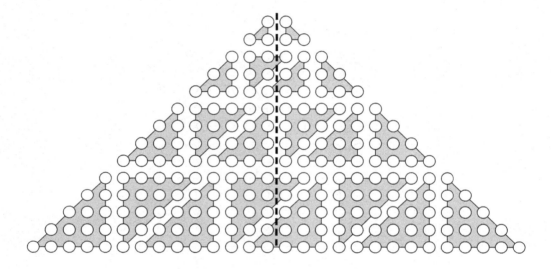

$$2\left[t_2 + 2t_3 + 3t_4 + 4t_5\right] = 2t_{14} = 2t_{t_5-1}.$$

Corollary.

$$\sum_{k=1}^{n} \binom{k+2}{3} = \binom{n+3}{4}.$$

—RBN

Centered Triangular Numbers

The *centered triangular number* c_n enumerates the number of dots in an array with one central dot surrounded by dots in n triangular borders, as illustrated below for $c_0 = 1$, $c_1 = 4$, $c_2 = 10$, $c_3 = 19$, $c_4 = 31$, and $c_5 = 46$:

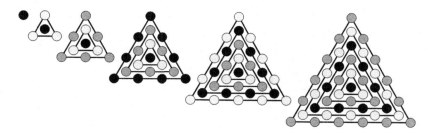

The ordinary triangular number t_n is equal to $1 + 2 + 3 + \cdots + n$.

I. Every $c_n \geq 4$ is one more than three times an ordinary triangular number, i.e., $c_n = 1 + 3t_n$ for $n \geq 1$.

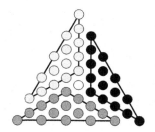

$$c_5 = 46 = 1 + 3 \cdot 15 = 1 + 3(1 + 2 + 3 + 4 + 5) = 1 + 3t_5.$$

II. Every $c_n \geq 10$ is the sum of three consecutive ordinary triangular numbers, i.e., $c_n = t_{n-1} + t_n + t_{n+1}$ for $n \geq 2$.

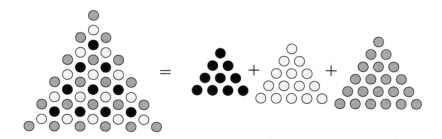

$$c_5 = 46 = 10 + 15 + 21 = t_4 + t_5 + t_6.$$

Jacobsthal Numbers

(Ernst Erich Jacobsthal, 1882–1965)

Let a_n be the number of ways of tiling a $3 \times n$ rectangle with 1×1 and 2×2 squares; b_n be the number of ways of filling a $2 \times 2 \times n$ hole with $1 \times 2 \times 2$ bricks, and c_n be the number of ways of tiling a $2 \times n$ rectangle with 1×2 rectangles and 2×2 squares. Then for all $n \geq 1$,

$$a_n = b_n = c_n.$$

Proof.

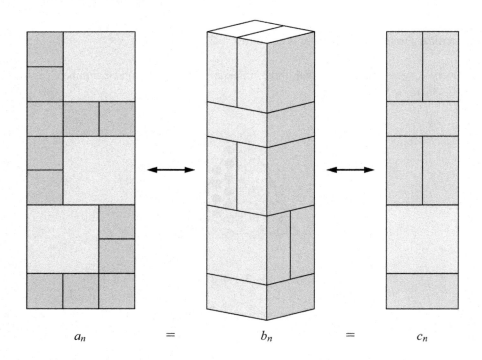

a_n $\qquad$ = $\qquad$ b_n $\qquad$ = $\qquad$ c_n

NOTE. $\{a_n\}_{n=1}^{\infty} = \{1, 3, 5, 11, 21, 43, \cdots\}$. These are the *Jacobsthal numbers*, sequence A001045 in the *On-Line Encyclopedia of Integer Sequences* at http://oeis.org.

—Martin Griffiths

Infinite Series
&
Other Topics

Geometric Series V

I. $\frac{1}{3} + \left(\frac{1}{3}\right)^2 + \left(\frac{1}{3}\right)^3 + \cdots = \frac{1}{2}$:

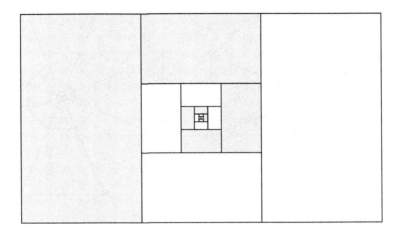

II. $\frac{1}{5} + \left(\frac{1}{5}\right)^2 + \left(\frac{1}{5}\right)^3 + \cdots = \frac{1}{4}$:

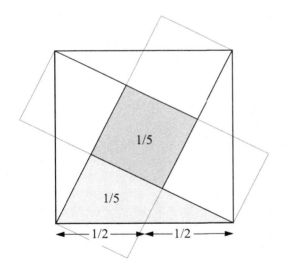

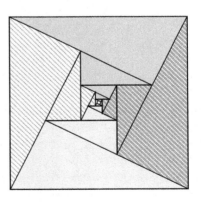

—Rick Mabry

Geometric Series VI

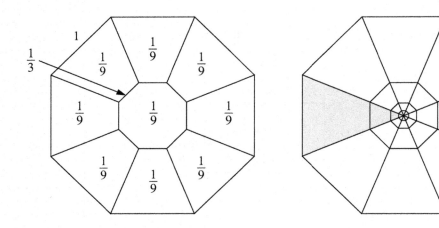

$$\frac{1}{9} + \frac{1}{9^2} + \frac{1}{9^3} + \cdots = \frac{1}{8}.$$

The general result $\frac{1}{n} + \frac{1}{n^2} + \frac{1}{n^3} + \cdots = \frac{1}{n-1}$ can be proved using this construction with a regular $(n-1)$-gon (or even a circle).

—James Tanton

Geometric Series VII (via Right Triangles)

$$\frac{1}{2} + \left(\frac{1}{2}\right)^2 + \left(\frac{1}{2}\right)^3 + \cdots = 1.$$

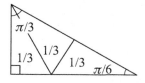 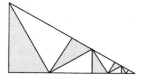

$$\frac{1}{3} + \left(\frac{1}{3}\right)^2 + \left(\frac{1}{3}\right)^3 + \cdots = \frac{1}{2}.$$

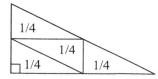 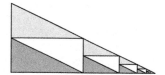

$$\frac{1}{4} + \left(\frac{1}{4}\right)^2 + \left(\frac{1}{4}\right)^3 + \cdots = \frac{1}{3}.$$

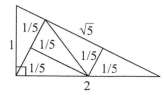 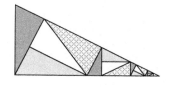

$$\frac{1}{5} + \left(\frac{1}{5}\right)^2 + \left(\frac{1}{5}\right)^3 + \cdots = \frac{1}{4}.$$

Challenge. Can you create the next two rows?

—RBN

Geometric Series VIII

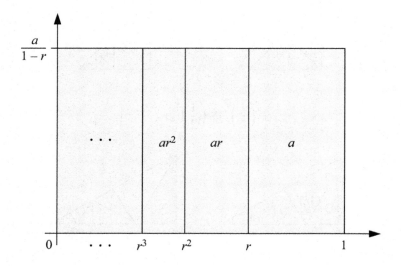

$$a > 0, \ r \in (0, 1) \quad \Rightarrow \quad a + ar + ar^2 + ar^3 + \cdots = \frac{a}{1 - r}.$$

—Craig M. Johnson & Carlos G. Spaht (independently)

Geometric Series IX

I. $a + ar + ar^2 + \cdots = \frac{a}{1-r}$, $0 < r < 1$:

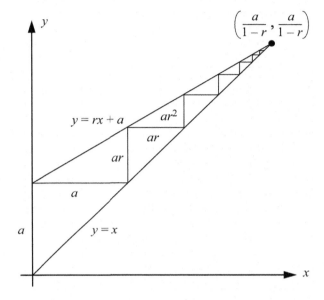

II. $a - ar + ar^2 - \cdots = \frac{a}{1+r}$, $0 < r < 1$:

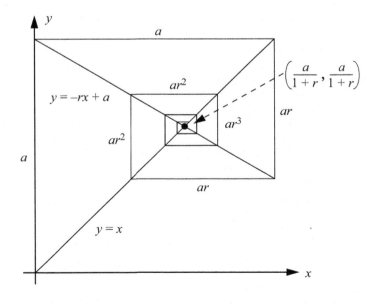

—The Viewpoints 2000 Group

Differentiated Geometric Series II

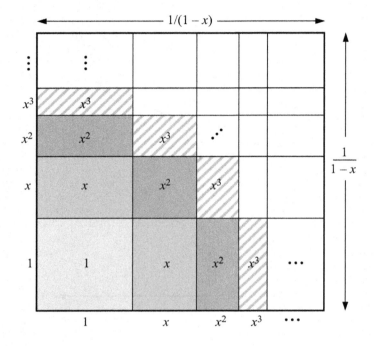

$$x \in [0, 1) \quad \Rightarrow \quad 1 + 2x + 3x^2 + 4x^3 + \cdots = \left(\frac{1}{1-x}\right)^2$$

—RBN

A Geometric Telescope

The two most basic series whose sums can be computed explicitly (geometric series, telescoping series) combine forces to demonstrate the amusing fact that

$$\sum_{m=2}^{\infty} (\zeta(m) - 1) = 1,$$

where $\zeta(s) = \sum_{n=1}^{\infty} \frac{1}{n^s}$ is the Riemann zeta function. Namely,

$\zeta(2)-1$	$\zeta(3)-1$	$\zeta(4)-1$	$\cdots$	
$\downarrow$	$\downarrow$	$\downarrow$		
$\dfrac{1}{2^2}$	$\dfrac{1}{2^3}$	$\dfrac{1}{2^4}$	$\cdots$	$= \dfrac{1/2^2}{1-1/2} = \dfrac{1}{2^2} \cdot \dfrac{2}{1} = \dfrac{1}{2\cdot 1} = 1 - \dfrac{1}{2}$
$\dfrac{1}{3^2}$	$\dfrac{1}{3^3}$	$\dfrac{1}{3^4}$	$\cdots$	$= \dfrac{1/3^2}{1-1/3} = \dfrac{1}{3^2} \cdot \dfrac{3}{2} = \dfrac{1}{3\cdot 2} = \dfrac{1}{2} - \dfrac{1}{3}$
$\dfrac{1}{4^2}$	$\dfrac{1}{4^3}$	$\dfrac{1}{4^4}$	$\cdots$	$= \dfrac{1/4^2}{1-1/4} = \dfrac{1}{4^2} \cdot \dfrac{4}{3} = \dfrac{1}{4\cdot 3} = \dfrac{1}{3} - \dfrac{1}{4}$
$\vdots$	$\vdots$	$\vdots$	$\ddots$	$= \qquad\qquad \cdots \qquad\qquad = \cdots$

$$\underline{}$$
$$1$$

Exercises. (a) $\sum_{m=2}^{\infty} (-1)^m (\zeta(m) - 1) = \frac{1}{2}$; (b) $\sum_{k=1}^{\infty} (\zeta(2k+1) - 1) = \frac{1}{4}$.

—Thomas Walker

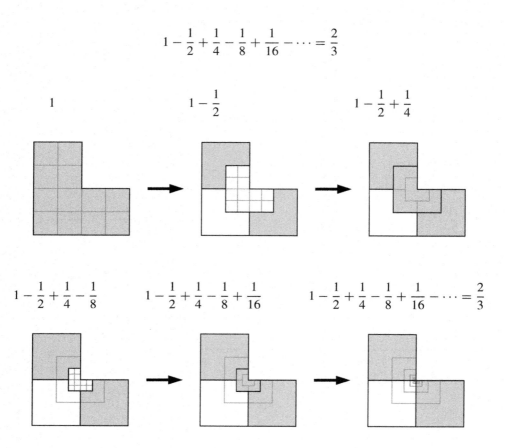

An Alternating Series II

$$1 - \frac{1}{2} + \frac{1}{4} - \frac{1}{8} + \frac{1}{16} - \cdots = \frac{2}{3}$$

1

$1 - \dfrac{1}{2}$

$1 - \dfrac{1}{2} + \dfrac{1}{4}$

$1 - \dfrac{1}{2} + \dfrac{1}{4} - \dfrac{1}{8}$

$1 - \dfrac{1}{2} + \dfrac{1}{4} - \dfrac{1}{8} + \dfrac{1}{16}$

$1 - \dfrac{1}{2} + \dfrac{1}{4} - \dfrac{1}{8} + \dfrac{1}{16} - \cdots = \dfrac{2}{3}$

—RBN

An Alternating Series III

$$1 - \frac{1}{3} + \frac{1}{9} - \frac{1}{27} + \frac{1}{81} - \cdots = \frac{3}{4}$$

1

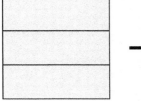

$1 - \dfrac{1}{3}$

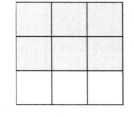

$1 - \dfrac{1}{3} + \dfrac{1}{9}$

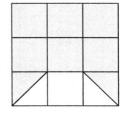

$1 - \dfrac{1}{3} + \dfrac{1}{9} - \dfrac{1}{27}$

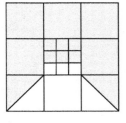

$1 - \dfrac{1}{3} + \dfrac{1}{9} - \dfrac{1}{27} + \dfrac{1}{81}$

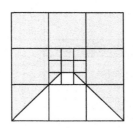

$1 - \dfrac{1}{3} + \dfrac{1}{9} - \dfrac{1}{27} + \cdots = \dfrac{3}{4}$

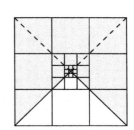

—Hasan Unal

The Alternating Series Test

Theorem. An alternating series $a_1 - a_2 + a_3 - a_4 + a_5 - a_6 + \cdots$ converges to a sum S if $a_1 \geq a_2 \geq a_3 \geq a_4 \geq \cdots \geq 0$ and $a_n \to 0$. Moreover, if $s_n = a_1 - a_2 + a_3 - \cdots + (-1)^{n+1} a_n$ is the nth partial sum, then $s_{2n} < S < s_{2n+1}$.

Proof.

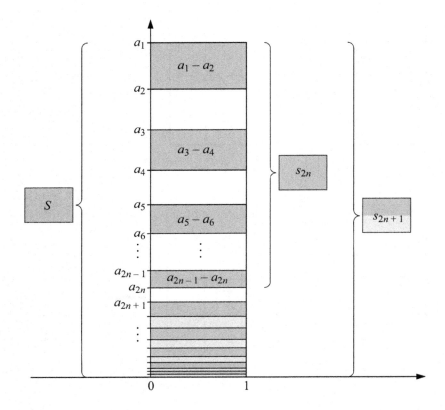

—Richard Hammack & David Lyons

The Alternating Harmonic Series II

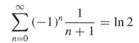

$$\sum_{n=0}^{\infty} (-1)^n \frac{1}{n+1} = \ln 2$$

$1 \cdots$

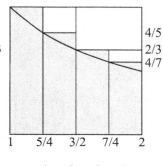

$-\dfrac{1}{2} + \dfrac{1}{3} \cdots$

$-\dfrac{1}{4} + \dfrac{1}{5} - \dfrac{1}{6} + \dfrac{1}{7} \cdots$

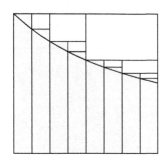

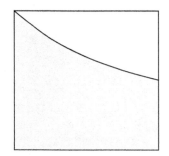

$-\dfrac{1}{8} + \dfrac{1}{9} - \dfrac{1}{10} + \dfrac{1}{11} - \dfrac{1}{12} + \dfrac{1}{13} - \dfrac{1}{14} + \dfrac{1}{15} \cdots$

$= \displaystyle\int_{1}^{2} \frac{1}{x}\,dx = \ln 2.$

—Matt Hudelson

Galileo's Ratios II

(Galileo Galilei, 1564–1642)

$$\frac{1}{3} = \frac{1+3}{5+7} = \frac{1+3+5}{7+9+11} = \cdots = \frac{1+3+\cdots+(2n-1)}{(2n+1)+(2n+3)+\cdots+(4n-1)}$$

$$= \frac{n^2}{(2n)^2 - n^2} = \frac{n^2}{3n^2} = \frac{1}{3}$$

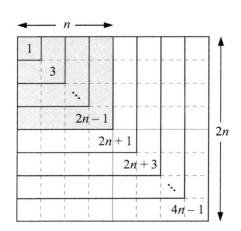

—Alfinio Flores & Hugh A. Sanders

Slicing Kites Into Circular Sectors

Areas: $\displaystyle\sum_{n=1}^{\infty} \frac{2^n\left[1-\cos(x/2^n)\right]^2}{\sin(x/2^{n-1})} = \tan\left(\frac{x}{2}\right) - \frac{x}{2}, \quad |x| < \pi$

Side Lengths: $\displaystyle 2\sum_{n=1}^{\infty} \frac{1-\cos(x/2^n)}{\sin(x/2^{n-1})} = \tan\left(\frac{x}{2}\right), \quad |x| < \pi$

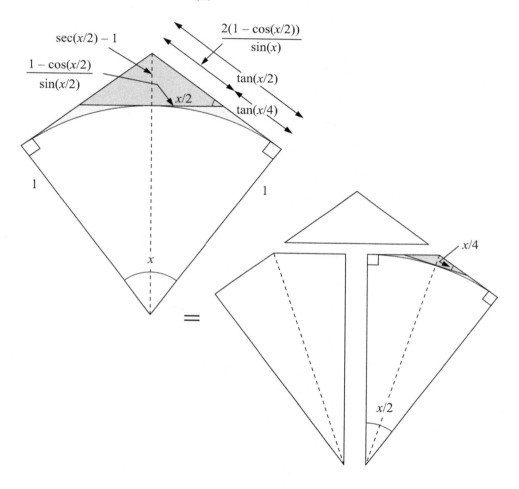

—Marc Chamberland

Nonnegative Integer Solutions and Triangular Numbers

For i, j, and k integers between 0 and n inclusive, the number of nonnegative integer solutions of $x + y + z = n$ with $x \le i$, $y \le j$, and $z \le k$ is

$$t_{i+j+k-n+1} - t_{j+k-n} - t_{i+k-n} - t_{i+j-n},$$

where $t_m = 1 + 2 + \cdots + m$ is the mth triangular number for $m \ge 1$ and $t_m = 0$ for $m \le 0$. E.g., $(n, i, j, k) = (23, 15, 11, 17)$:

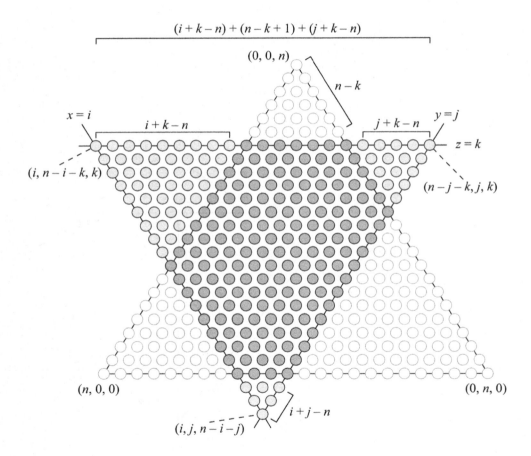

—Matthew J. Haines & Michael A. Jones

Dividing a Cake

To cut a frosted rectangular cake into n pieces so that each person gets the same amount of cake and frosting:

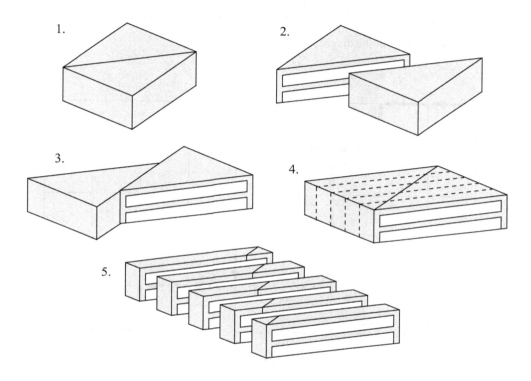

1.

2.

3.

4.

5.

—Nicholaus Sanford

The Number of Unordered Selections with Repetitions

Theorem. The number of unordered selections of r objects chosen from n types with repetitions allowed is $\dbinom{n-1+r}{r}$, the same as the number of paths of length $n-1+r$ from top-left to lower-right in the diagram.

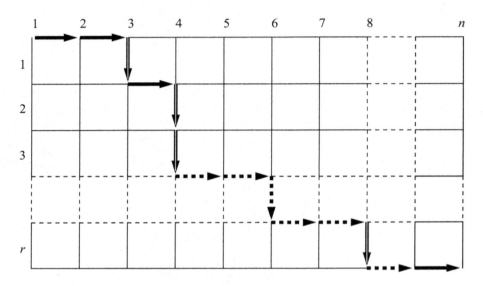

Selection $3, 4, 4, \ldots, 6, \ldots, 8$.

—Derek Christie

A Putnam Proof Without Words

(Problem A1, 65th Annual William Lowell Putnam Mathematical Competition, 2004)

Basketball star Shanille O'Keal's team statistician keeps track of the number $S(N)$ of successful free throws she has made in her first N attempts of the season. Early in the season $S(N)$ was less than 80% of N, but by the end of the season, $S(N)$ was more than 80% of N. Was there necessarily a moment in between when $S(N)$ was *exactly* 80% of N?

Answer. Yes.

Proof.

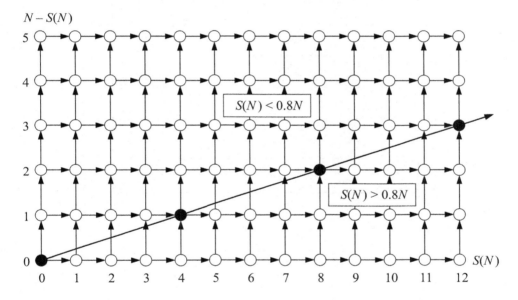

Exercises. (a) Answer the same question assuming that Shanille had $S(N) > 0.8N$ early in the season and $S(N) < 0.8N$ at the end; (b) What other values could be substituted for 80% in the original question?

—Robert J. MacG. Dawson

On Pythagorean Triples

Theorem. There exists a one-to-one correspondence between Pythagorean triples and factorizations of even squares of the form $n^2 = 2pq$.

Proof by inclusion-exclusion, e.g., for $6^2 = 2 \cdot 2 \cdot 9$:

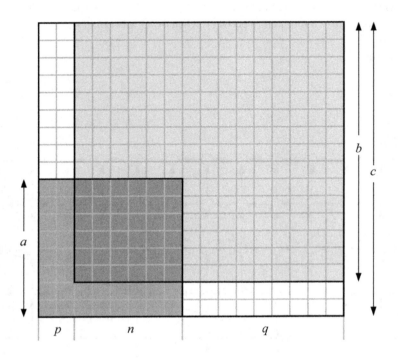

$$c^2 = a^2 + b^2 - n^2 + 2pq,$$
$$\therefore c^2 = a^2 + b^2 \Leftrightarrow n^2 = 2pq.$$

—José A. Gomez

Pythagorean Quadruples

A *Pythagorean quadruple* (a, b, c, d) of positive integers satisfies $a^2 + b^2 + c^2 = d^2$. A formula that generates infinitely many Pythagorean quadruples is

$$(m^2 + p^2 - n^2)^2 + (2mn)^2 + (2pn)^2 = (m^2 + p^2 + n^2)^2.$$

Proof.

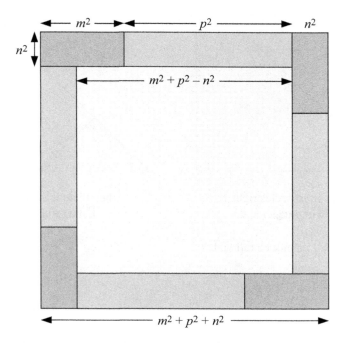

NOTE. While the formula generates infinitely many Pythagorean quadruples, it does not generate all of them, e.g., it does not generate (2,3,6,7). A formula that does generate all Pythagorean quadruples is

$$(m^2 + n^2 - p^2 - q^2)^2 + (2mq + 2np)^2 + (2nq - 2mp)^2 = (m^2 + n^2 + p^2 + q^2)^2.$$

—RBN

The Irrationality of $\sqrt{2}$

By the Pythagorean theorem, an isosceles triangle of edge length 1 has hypotenuse $\sqrt{2}$. If $\sqrt{2}$ is rational, then some positive integer multiple of this triangle must have three sides with integer lengths, and hence there must be a *smallest* isosceles right triangle with this property. However,

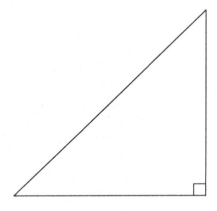

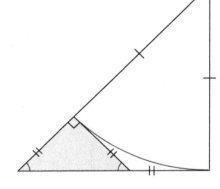

if this is an isosceles right
triangle with integer sides,

then there is a smaller one
with the same property.

Therefore $\sqrt{2}$ cannot be rational.

—Tom M. Apostol

$\mathbb{Z} \times \mathbb{Z}$ is a Countable Set

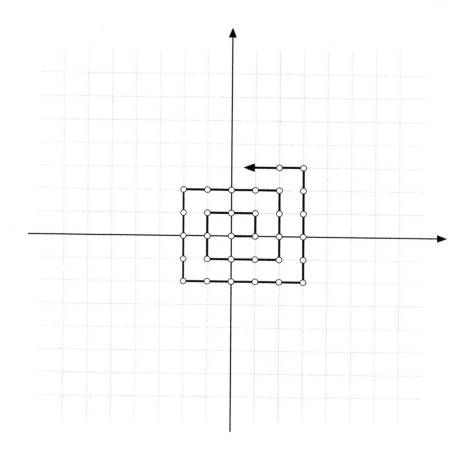

—Des MacHale

A Graph Theoretic Summation of the First n Integers

$$\sum_{i=1}^{n} i = \binom{n+1}{2}$$

E.g., $n = 5$.

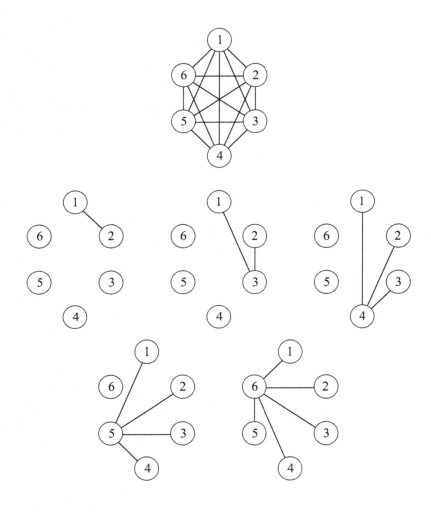

—Joe DeMaio & Joey Tyson

A Graph Theoretic Decomposition of Binomial Coefficients

$$\binom{n+m}{2} = \binom{n}{2} + \binom{m}{2} + nm$$

E.g., $n = 5$, $m = 3$.

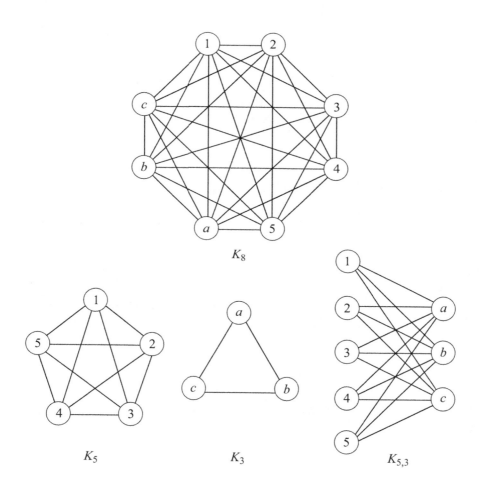

K_8

K_5 K_3 $K_{5,3}$

—Joe DeMaio

(0,1) and [0,1] Have the Same Cardinality

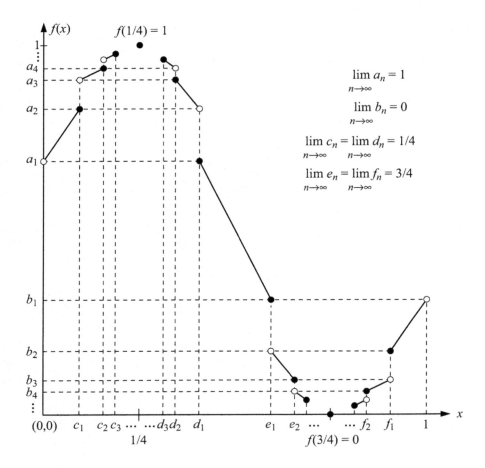

$$\lim_{n\to\infty} a_n = 1$$

$$\lim_{n\to\infty} b_n = 0$$

$$\lim_{n\to\infty} c_n = \lim_{n\to\infty} d_n = 1/4$$

$$\lim_{n\to\infty} e_n = \lim_{n\to\infty} f_n = 3/4$$

—Kevin Hughes & Todd K. Pelletier

A Fixed Point Theorem

One of the best pictorial arguments is a proof of the "fixed point theorem" in one dimension: *Let $f(x)$ be continuous and increasing in $0 \leq x \leq 1$, with values satisfying $0 \leq f(x) \leq 1$, and let $f_2(x) = f(f(x))$, $f_n(x) = f(f_{n-1}(x))$. Then under iteration of f every point is either a fixed point, or else converges to a fixed point.*

For the professional the only proof needed is [the figure]:

A Mathematician's Miscellany (1953)

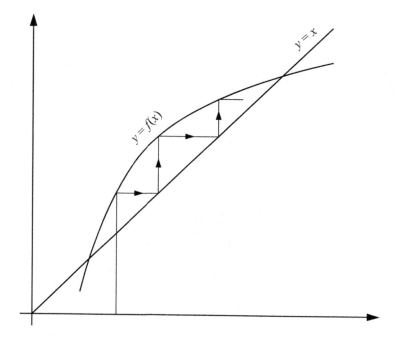

—John Edensor Littlewood

In Space, Four Colors are not Enough

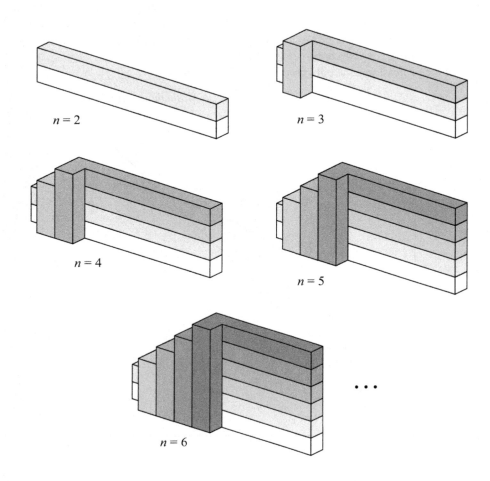

$n = 2$

$n = 3$

$n = 4$

$n = 5$

$n = 6$

$\cdots$

—Claudi Alsina & RBN

Sources

page   source

Geometry & Algebra (continued)

Trigonometry, Calculus & Analytic Geometry

page   source

Integers & Integer Sums (continued)

115   *Mathematics Magazine*, vol. 78, no. 5 (Dec. 2005), p. 385.
116   *Mathematical Intelligencer*, vol. 22, no. 3 (Summer 2000), p. 47–49.
117   *College Mathematics Journal*, vol. 31, no. 5 (Nov. 2000), p. 392.
118   *College Mathematics Journal*, vol. 45, no. 1 (Jan. 2014), p. 16.
119   *Mathematics Magazine*, vol. 80, no. 1 (Feb. 2007), pp. 74–75.
120   *Mathematics Magazine*, vol. 74, no. 4 (Oct. 2001), pp. 314–315.
121   *College Mathematics Journal*, vol. 45, no. 5 (Nov. 2014), p. 349.
122   *College Mathematics Journal*, vol. 44, no. 4 (Sept. 2013), p. 283.
123   *Mathematical Intelligencer*, vol. 24, no. 4 (Fall 2002), pp. 67–69.
124   *College Mathematics Journal*, vol. 33, no. 2 (March 2002), p. 171.
125   *Mathematics Magazine*, vol. 76, no. 2 (April 2003), p. 136.
126   *Mathematics Magazine*, vol. 85, no. 5 (Dec. 2012), p. 360.
127   *College Mathematics Journal*, vol. 45, no. 2 (March 2014), p. 135.
128   *Charming Proofs*, MAA, 2010, pp. 18, 240.
129   *Mathematics Magazine*, vol. 81, no. 4 (Oct. 2008), p. 302.
130   *Mathematics Magazine*, vol. 84, no. 4 (Oct. 2011), p. 295.
131   *Mathematics Magazine*, vol. 86, no. 1 (Feb. 2013), p. 55.
132   I. *Math Made Visual*, MAA, 2006, pp. 18, 147.
        II. *Charming Proofs*, MAA, 2010, p. 14.
133   *Mathematics Magazine*, vol. 77, no. 3 (June 2004), p. 200.
134   *College Mathematics Journal*, vol. 40, no. 2 (March 2009), p. 86.
135   *Mathematics and Computer Education*, vol. 31, no. 2 (Spring 1997), p. 190.
136   *Mathematics Magazine*, vol. 77, no. 5 (Dec. 2004), p. 373.
137   *Mathematics Magazine*, vol. 79, no. 1 (Feb. 2006), p. 44.
138   *Mathematics Magazine*, vol. 78, no. 3 (June 2005), p. 231.
139   *Mathematics Magazine*, vol. 80, no. 1 (Feb. 2007), p. 76.
140   *Mathematics Magazine*, vol. 85, no. 5 (Dec. 2012), p. 373.
141   *College Mathematics Journal*, vol. 46, no. 2 (March 2015), p. 98.
142   *College Mathematics Journal*, vol. 44, no. 3 (May 2013), p. 189.
143   *College Mathematics Journal*, vol. 16, no. 5 (Nov. 1985), p. 375.
144   *Mathematics Magazine*, vol. 79, no. 1 (Feb. 2006), p. 65.
145   *College Mathematics Journal*, vol. 34, no. 4 (Sept. 2003), p. 295.
146   *Mathematics Magazine*, vol. 77, no. 5 (Dec. 2004), p. 395.
147   *Mathematics Magazine*, vol. 78, no. 5 (Dec. 2005), p. 395.
148   *Mathematics Magazine*, vol. 79, no. 4 (Oct. 2006), p. 317.
150   *College Mathematics Journal*, vol. 41, no. 2 (March 2010), p. 100.

Infinite Series and Other Topics

153   I. *College Mathematics Journal*, vol. 32, no. 1 (Jan. 2001), p. 19.
        II. http://lsusmath.rickmabry.org/rmabry/fivesquares/fsq2.gif
154   *College Mathematics Journal*, vol. 39, no. 2 (March 2008), p. 106.

NOTE: Several of the PWWs in this book (pp. 4, 35, 63, 79, and 100) are not listed here as they may not have previously appeared in print.

Index of Names

About the Author

Roger B. Nelsen was born in Chicago, Illinois. He received his B.A. in mathematics from DePauw University in 1964 and his Ph.D. in mathematics from Duke University in 1969. Roger was elected to Phi Beta Kappa and Sigma Xi, and taught mathematics and statistics at Lewis & Clark College for forty years before his retirement in 2009. His previous books include *Proofs Without Words*, MAA 1993; *An Introduction to Copulas*, Springer, 1999 (2nd ed. 2006); *Proofs Without Words II*, MAA, 2000; *Math Made Visual* (with Claudi Alsina), MAA, 2006; *When Less Is More* (with Claudi Alsina), MAA, 2009; *Charming Proofs* (with Claudi Alsina), MAA, 2010; *The Calculus Collection* (with Caren Diefenderfer), MAA, 2010; *Icons of Mathematics* (with Claudi Alsina), MAA, 2011, *College Calculus* (with Michael Boardman), MAA, 2015, *A Mathematical Space Odyssey* (with Claudi Alsina), MAA, 2015, and *Cameos for Calculus*, MAA 2015.